SPECTER STARGAZER

Written and Illustrated By: Jessica Specter

For more information, contact:
specterbyjessica@outlook.com
ISBN: 979-8-3303-8569-0

Hello Stargazer,

Within this work you will find a dazzling way to reimagine the cosmos and spirituality. I invite everyone to utilize this book how it best suits them, as I know many people from varying interests and backgrounds will hold this - astronomers, space fanatics, spiritual practitioners, students and more!

In all, I hope you find this to be entertaining, insightful and fun. After completing this book you will be able to gaze upon the night sky and tell amazing stories of the stars to the people you are with. To read the constellations and retell mythological correspondences is a lost art.

Not only that - you will be learning facts about space, the stars in some of the constellations, seasons and more. If you practice divination, your readings will have a new edge and greater depth than ever before.

- Jessica Specter

Table of Contents

The Planets

MERCURY

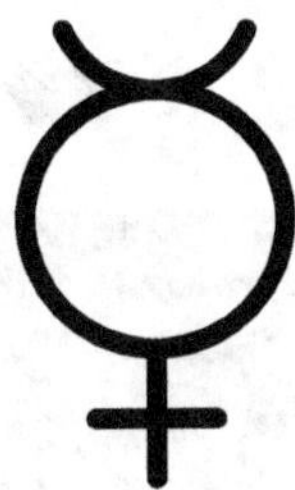

Mercury (also recognized as Hermes) was the messenger god and mediator between the realm of the dead and the living. He would guide souls to the Underworld. He is the god of financial gain, commerce, eloquence, messages, communication (including divination), travelers, boundaries, luck, trickery, and thieves

In astrology, Mercury is about quick ideation, reasoning and rationale in an energizing way. Mercury can also help in expression.

ELEMENT: Earth

ZODIAC: Gemini and Virgo

NUMEROLOGY: 5

GENERAL: Communication, intellect, trickery, memory and transportation.

Mercury

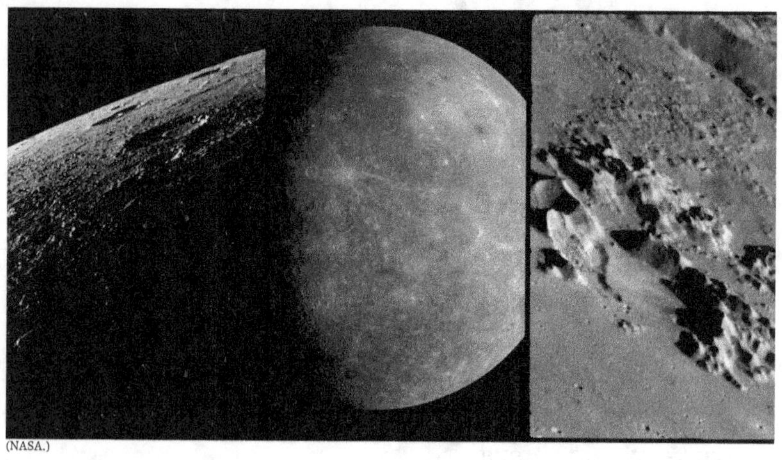

This planet is slightly larger than Earth's moon - and it looks similar. It has a thin atmosphere. It's magnetic field is weaker than Earth's; suggesting a molten core. It has an oval orbit.

Mercury was named after Mercurius/Hermes because of it's *swift* movement through the sky and in orbit.

Year Length: 88 days

Mercurius/Hermes

Hermes is the son of Zeus. He was born in a cave on Mount Cyllene in Arcadia. The day he was born, he escaped his crate and stole Apollo's cattle. He tricked Apollo via his lyre, so much so, that Apollo traded the cattle for the lyre. This established the young God not only as a trickster but an inventor.

Hermes duties were to relay messages between the Gods and guide souls to the Underworld. He is seen as the boundary between realms because his speed aided in his job as an intermediary between im/mortal worlds, thus bringing balance.

- - -

Other myths:

Pandora: Hermes bestowed the first woman (Pandora) with speech and cunningness. This attributed to her later bestowing blessings and curses via opening the fateful jar.

Odysseus: Hermes gave him "Moly" to avoid Circe's charm.

In other stories, he usually acts as a guide, protector and negotiator. He can be both helpful and mischievous.

VENUS

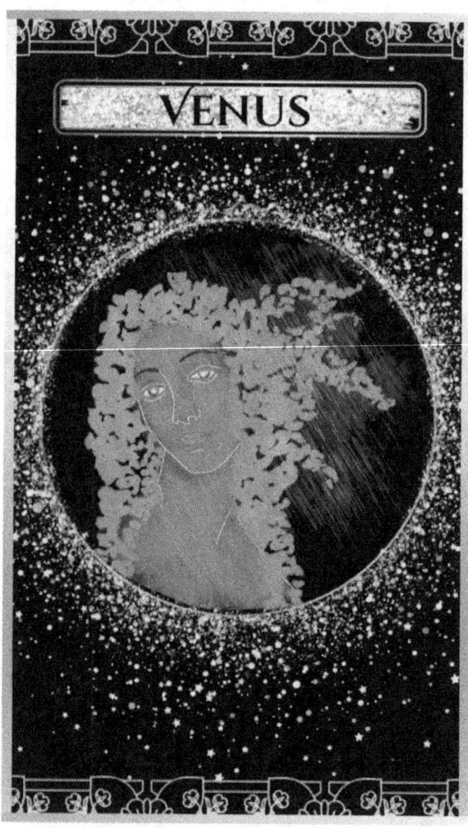

Aphrodite was born from the white foam produced by the severed genitals of Uranus, the personification of heaven, after his son Kronos threw them into the sea.

Your natal Venus speaks to the way you express your desires, your passions, what you value, and how you relate to and experience pleasure. It also influences how you socialize, relate to, and attract others.

ELEMENT: Water

ZODIAC: Taurus and Libra

NUMEROLOGY: 6

GENERAL: Love, romance, beauty, fertility and art.

Venus

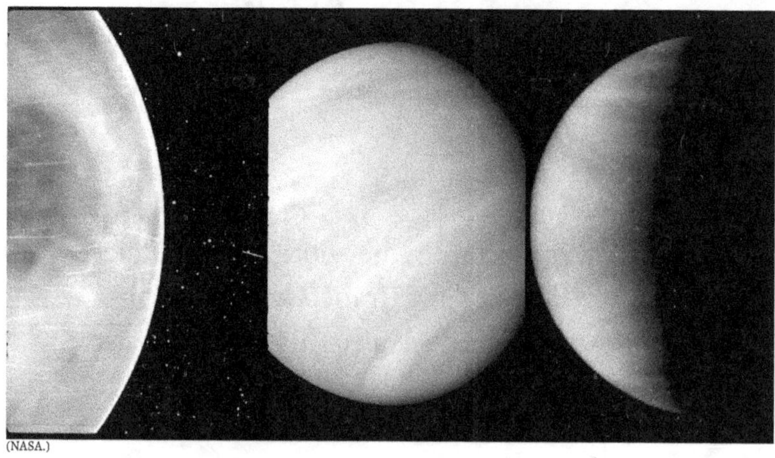

(NASA.)

Called Earth's "evil twin." The hottest planet in the solar system. The sulfur-filled (rotten egg smell) atmosphere is so thick that the Sun looks like a *smear* of light. Venus rotates backwards on it's axis - meaning that the Sun rises and sets opposite as ours does.

Year Length: 243 days

Venus/Aphrodite

Aphrodite was birthed when Kronos severed Uranus' genitals (his father) into the sea. She is recognized as the goddess of love and beauty. She married Hephaestus, the God of fire and blacksmiths, due to Zeus' bidding to diminish conflict between the Gods - there was much tension and infidelity (with Ares and Adonis.) Aphrodite had 3 children with Ares.

Symbols: roses, swans, doves and myrtle.

- - -

Other myths:

Judgement of Paris: Paris, the Prince of Troy (think of the horse), was asked to judge who was the fairest between three goddesses. Each one bribed him and he chose Aphrodite who offered him Helena, the most beautiful woman of Sparta. This indirectly led to the Trojan war, as Helena was already married to the King of Sparta (be careful what you wish for.)

Pygmalion & Galatea: Pygmalion fell in love with a statue that he carved of a woman. Aphrodite took pity on him and brought her to life. Galatea became his wife.

MARS

Mars is known as the god of war, son of Jupiter and Juno. His corresponding animals are the wolf and the woodpecker.

Mars, "The Red Planet" is the planet of action. Mars commands you to get up and get things done. He challenges you to do the best that you can. Too much of this energy could also result in destruction, so it is important to balance.

ELEMENT: Fire

ZODIAC: Aries

NUMEROLOGY: 9

GENERAL: Energy, passion, action and desire.

Mars

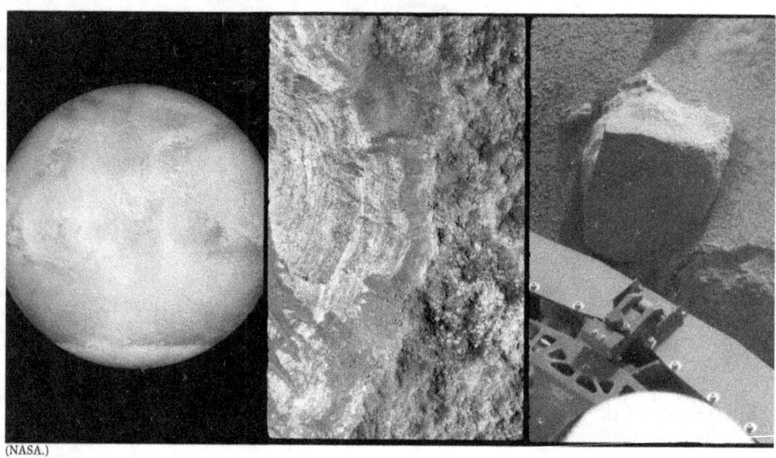

This planet is red due to the iron oxide that covers its surface. It is half the size of Earth and the average temperature is -85°F. Like our planet, there are four seasons but each one lasts twice as long due to it's orbit length.

There are large, planet-wide dust storms that occur. It has two moons, Phobos and Deimos. There is ice here, but no liquid water.

Year Length: 687 days

Mars/Ares

Mars and Ares are both gods of war. Ares is considered to be more chaotic, brutual in terms of conflict. Mars was seen more as a force of protection and military prowess.

Ares was born to Zeus and Hera. He was unliked by other Gods and was seen as blood thirsty. He had two children, Deimos (terror) and Phobos (fear.) For Mars, the month of March was named after him and during this time they celebrated the 'Feast of Mars' to invoke success in war.

Symbols: Shield, helmet, spear, chariot. The wolf and woodpecker are exclusive to Mars. Swords, dogs and vultures are exclusive to Ares.

- - -

Other myths:

Ares and the Trojan War: Initially Ares is on the side of the Trojans as he spread death and destruction. Athena strategically tricks him, and Diomedes wounds him causing him to flee back to Olympus.

Temple of Mars Ultor: Mars was closely associated to Augustus, the first Roman emperor who built Mars Ultor. The title recognizes his revenge for the assassination of Julius Caesar. It was a place where military decisions were made and where Mars would be invoked.

JUPITER

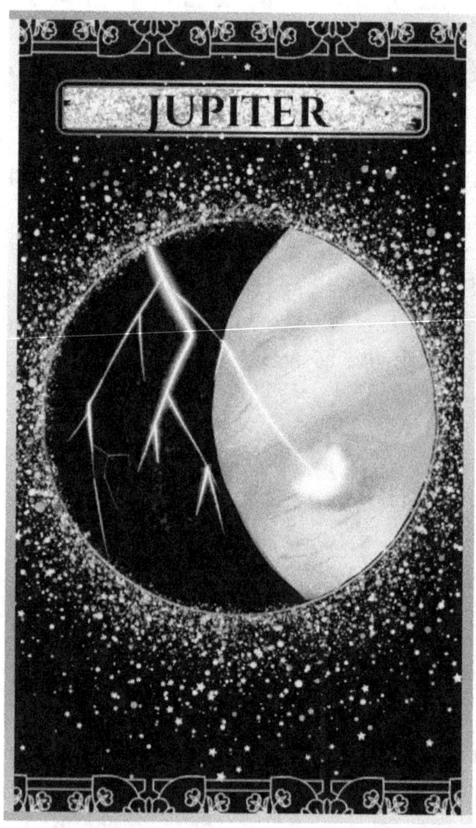

♃

Jupiter was the king of the gods. He was the god of the sky and thunder. His counterpart is known as Zeus.

Jupiter is widely referred to as "the Great Benefic," meaning it holds an abundance of gifts and is always prepared to bestow its generosity onto us. Jupiter's energy is joyous and optimistic, and it's often responsible for showing us how to remain positive in difficult times.

ELEMENT: Space

ZODIAC: Sagittarius

NUMEROLOGY: 3

GENERAL: Positivity, achievement, great luck, authority, influence, expansion and wealth.

Jupiter

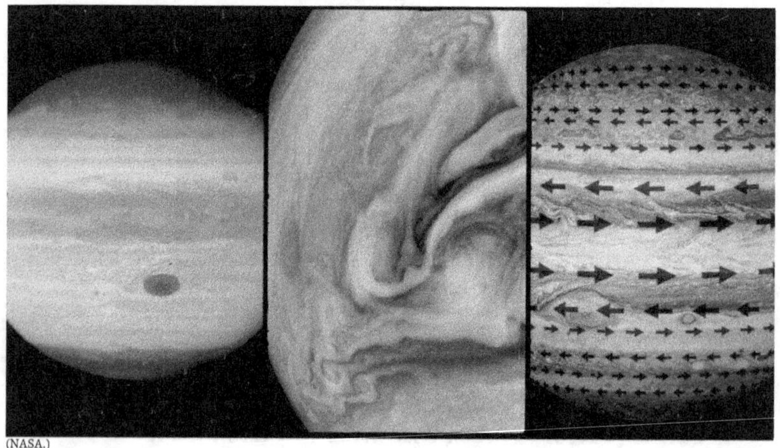

Jupiter is the largest planet within our solar system; more than twice as massive as all the other planets combined. Jupiter is a failed star - a gaseous planet with no solid surface. It has at least 79 moons with faint rings, likely caused by destruction of other moons.

One day on Jupiter lasts anywhere from 9 to 9 and a half hours. Its atmosphere is 600 miles thick. It's red spot is a humongous storm that's been raging for hundreds of years.

Year Length: 4,333 days, or 12 years.

Jupiter/Zeus

Jupiter is associated with the founders of Rome, Romulus and Remus, who were said to be the sons of Mars. Romulus sought Jupiter's favor to protect the city and its people, and in turn was worshipped as the supreme god. He was married to Juno/Hera and had frequent issues with their marriage as he practiced a lot of infidelity.

Symbols: Thunderbolt, eagle, oak tree and scepter. A bull and throne are exclusive to Zeus. A Capitol (temple) and wheel are exclusive to Jupiter.

- - -

Other myths:

Birth of Zeus: The youngest son of Kronos and Rhea who was hidden in a cave on the island of Crete to be saved from being swallowed by his father, like all of his other siblings were. He later released his siblings who would go on to fight the Titans and become the Olympian gods.

Zeus and the Flood: Angry with the wickedness of mankind, he flooded the earth to destroy humanity. Only Deucalion and his wife Pyrrha were warned by Prometheus and instructed to build an ark to survive. Zeus allows them to repopulate the earth by throwing stones behind them, which transform into men and women.

SATURN

Saturn is based on Kronos. Saturn urges us to deal with reality. Saturn, the Great Teacher planet, brings maturity and teaches us the value of patience and sacrifice.

Saturn is the god of time, generation, dissolution, abundance, wealth, agriculture, periodic renewal and liberation. Saturn's mythological reign was depicted as a Golden Age of abundance and peace.

ELEMENT: Earth

ZODIAC: Capricorn

NUMEROLOGY: 8

GENERAL: Time, matter, the boundary between above and below, the limitation of undefined things, bringing concepts to reality, spirit to body, thoughts to words, life to death.

Saturn

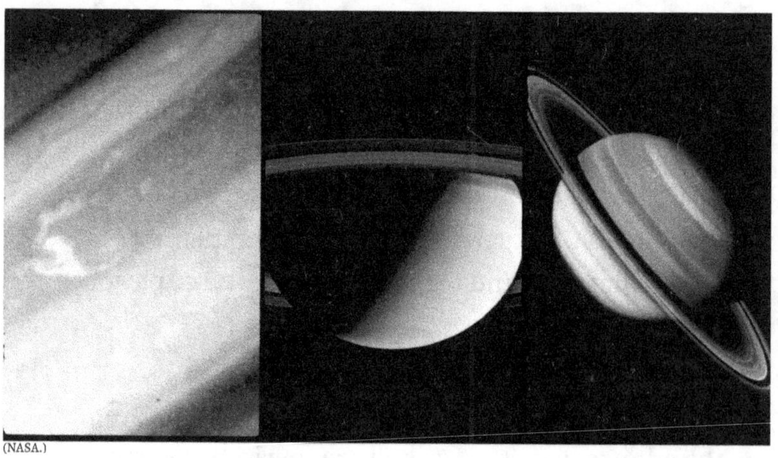

(NASA.)

Saturn's rings are made up of particles from the size of dust to mountains and is the second-largest planet in the solar system. It's a gas giant and less dense than water - meaning it could hypothetically float in water. A day on this planet lasts 10.7 hours.

At the north pole it beholds a hexagonal storm. 146 moons are confirmed to belong to Saturn. Some of it's smaller moons "shepherd" the rings by using their gravity to shape the particles in them.

Year Length: 29.5 years.

Saturn/Kronos

Saturn/Kronos are Gods associated with time of harvest/agriculture, time and wealth. In Greek mythology, Kronos is the leader of the Titans and father of the Olympian gods. He is often associated with the destructive aspect of time, cyclesmyt and inevitability of fate. He was birthed to Uranus and Gaia, his siblings being Titans, Cyclopes and Hecatoncheires. There was a festival, Saturnalia, with feasts, role reversals, merrymaking, etc. to bring back a temporary return of Saturn's rule.

Symbols: Scythe/sickle.

- - -

Other myths:

The Golden Age: Under Kronos' rule the world experienced a Golden Age in which there was peace, harmony and prosperity. Humans lived in abundance without the need for labor or hardship. There were no wars or strife.

Fear of Prophecy: Kronos feared his children would overthrow him (a prophecy) and so he swallowed all of his children but one at birth. Rhea, his wife and sister, hid Zeus.

Imprisonment: Upon defeat Kronos was imprisoned in Tartarus, a deep abyss used as a dungeon of suffering within the Underworld. He guides souls as well for redemption.

URANUS

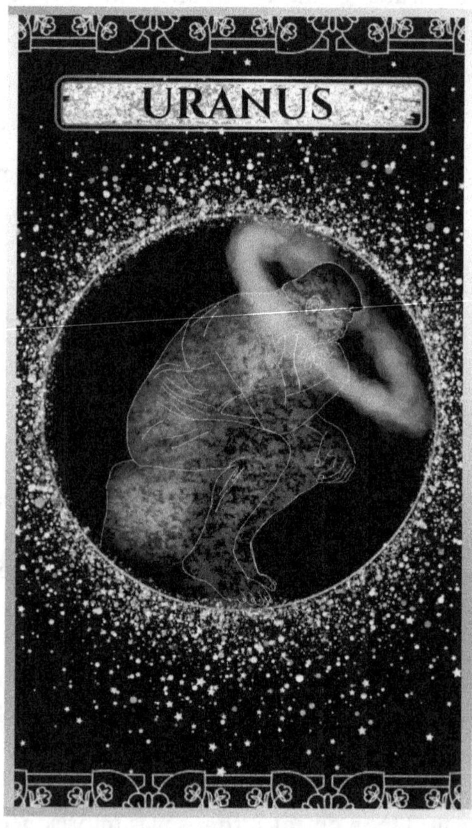

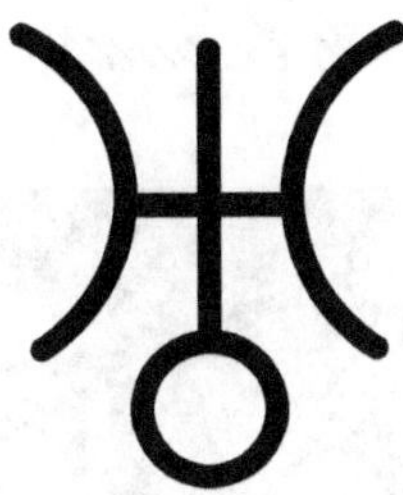

Uranus is the god of the heavens and sky and a symbol of masculinity. Uranus did not favor his children and locked them in Tartarus. Saturn (his son) would later steal Uranus' position of king of the Gods.

This revolutionary planet hates the rules and is always eager to facilitate groundbreaking, dynamic change. Uranus can have surprising effects.

ELEMENT: Air

ZODIAC: Aquarius

NUMEROLOGY: 4

GENERAL: Breakthrough, revolution, scientific genius and innovation.

Uranus

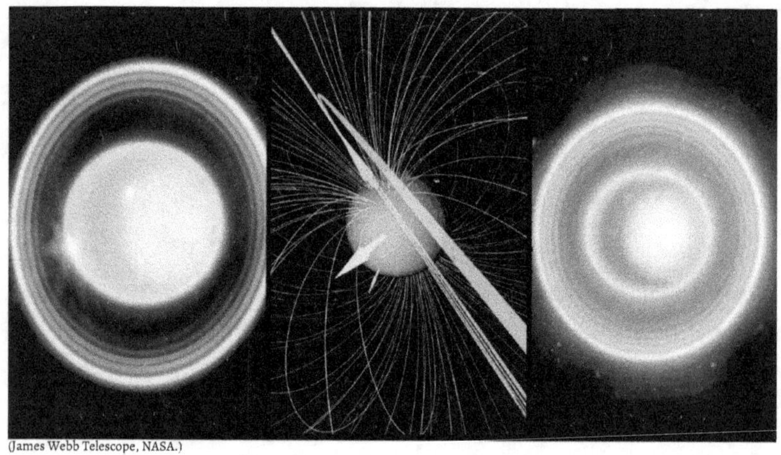

(James Webb Telescope, NASA.)

Uranus sits tilted on its side at around 98 degrees.
This causes extreme seasons with each pole
experiencing 42 years of daylight and 42 years of
darkness. It is also the coldest planet with a
minimum temperature of -371°F.

It is an ice giant with water, ammonia and methane
within its atmosphere - causing a blue-green color.
It holds 13 rings. It orbits the Sun in the opposite
direction of most planets; east to west.

Year Length: 84 years.

Uranus/Caelus

Uranus/Caelus are associated with the sky/heavens. He was born from Chaos/the void. Gaia (the Earth) is Uranus' wife. Uranus was a tyrannical father who imprisoned his children in Tartarus, distressing Gaia. He was castrated by his son Kronos with a sickle. The blood from this fell to the Earth and bore Furies and the Giants. This symbolized the overthrow of the old order and rise of the Titans.

Symbols: Arrow (Uranus.) Valut or dome (Caelus.)

- - -

Other myths:

Aphrodite's Birth: Upon castration into the sea, it created foam and from which birthed Aphrodite, the goddess of love and beauty.

Roman Caelus: Caelus is less prominent with the Romans. He is depicted as a bearded man with a cloud made of clouds, surrounded by the heavens. He symbolized the vastness and power of the sky, reflecting Roman ideals of strength and stability.

Curse: Kronos was cursed by Uranus - he decalred one of his own children would overthrow him.

NEPTUNE

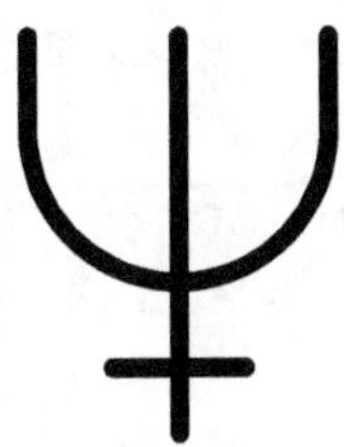

Neptune was the Roman god of waters and seas, who controlled winds and storms. Also known as Neptunus Equester, he was recognized as a god of horses and horsemanship, as well as patron of horse racing, a popular form of entertainment for the ancient Romans.

Neptune's glyph is the trident of Poseidon, God of the Seas. Much about this planet is fluid (Neptune rules the oceans of the Earth), changeable and illusory in nature.

ELEMENT: Water

ZODIAC: Pisces

NUMEROLOGY: 7

GENERAL: Dreams, illusion and abstract thought.

Neptune

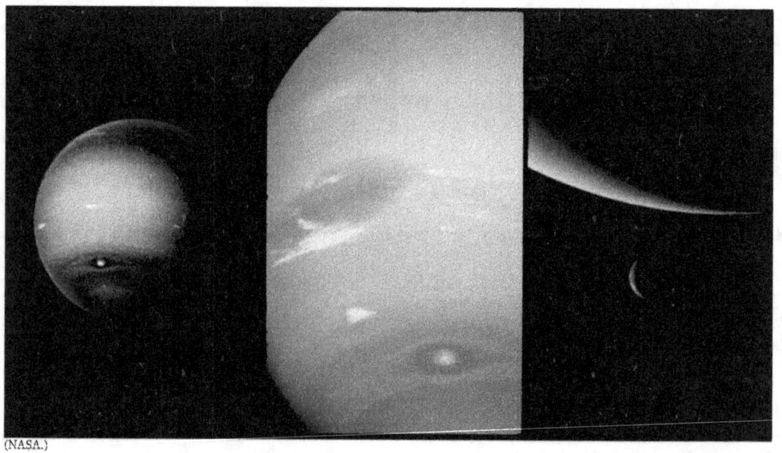

(NASA.)

Neptune is four times wider than Earth and has 16 hour days. It is mostly a mass of hot, dense fluid above a small and rocky core. It has icy materials such as water, methane and ammonia. It's atmosphere is mostly of hydrogen, atomic helium and methane.

14 moons orbit Neptune. It has rings, at least 5 main ones and four other ring arcs that are formed by the an orbiting moon's gravity. Pluto can sometimes be closer to the Sun than Neptune due to their orbit paths.

Year Length: 165 years.

Neptune/Poseidon

God of the sea, son of Kronos. His temperament is volatile.
When angered, he creates devastating storms, earthquakes
and shipwrecks. He can protect and destroy. Heavily
associated with the creation and majesty of horses.

Symbols: Trident, horses, water and sea creatures. Fish and
dolphins are exclusive to Poseidon. Shells are exclusive to
Neptune.

- - -

Other myths:

Contest for Athens: Athena and Poseidon fought for
patronage over Athens. Poseidon struck the ground with his
trident, creating a saltwater spring. Athena, offered the olive
tree which provided food, oil and wood. The citizens chose
Athena.

Trojan War: Initally supported the greeks. He sent Cetus to
attack Troy as punishment for not honoring him. He later
helped Troy build their walls but stopped when they didn't
honor him, once again.

Polyphemus: The cyclops, whom he is the father of, was
blinded by Odysseus to escape his cave. Poseidon seeks
revenge on Odysseus, making his journey home perilous.

PLUTO

Pluto was the lord of the subterranean underworld, which in Roman mythology served as the resting place of departed souls. He lived underground in a gloomy palace, and seemed to have little interest in the world of men. Likewise, Pluto seldom involved himself in godly affairs.

Pluto is also known as The Wealthy One, The Unseen and the Giver of Wealth.

ELEMENT: Water

ZODIAC: Scorpio

NUMEROLOGY: 0

GENERAL: Wealth, regeneration, rebirth and subconscious forces.

Pluto

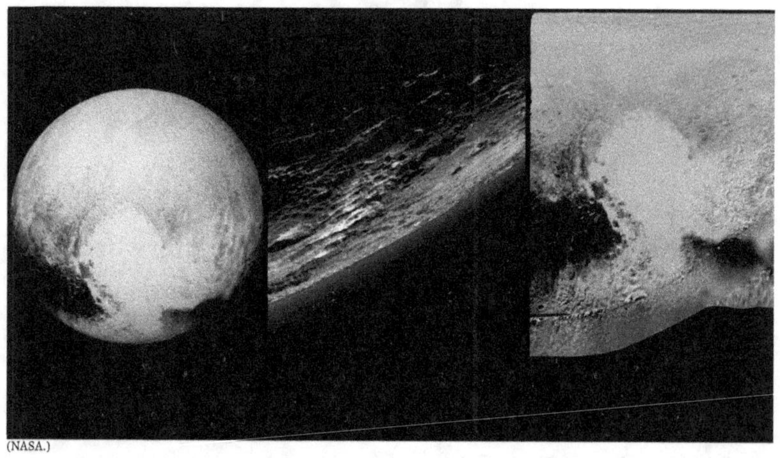

(NASA.)

Previously known as the 9th planet in our solar system, Pluto has been demoted as a "dwarf planet" in 2006. Pluto is smaller than Earth's Moon. Despite this, it has 5 of its own moons. It orbits around 3.7 billion miles away from the Sun. One day lasts 153 hours.

The largest moon beholden to Pluto is Charon. It also has Styx, Nix, Kerberos (Cerberus) and Hydra. Pluto's thin atmosphere expands and freezes when it gets closer to the Sun. It is a part of the Kuiper Belt, aka a region of icy objects beyond Neptune that has other dwarf planets, etc.

Year Length: 248 years.

Pluto/Hades

God of the Underworld and guardian of the dead, first-born son of Kronos.

Symbols: Cerberus, Scepter, Pomegranate, Chthonic symbols and riches. Pitchfork, Helm of Darkness, Gold and Wealth for Hades. Crown and laurel wreath for Pluto.

- - -

Persephone: the daughter of Demeter, Persephone, was abducted by Hades in his chariot when hesaw her gathering flowers to make her his queen. Demeter searches for her and neglects the earth, causing the seasons to change and turn barren. Zeus intervenes, and a compromise is reached. Persephone spends half the year with Hades and half with Demeter on Earth. This explains the changing of seasons.

Judgement of Souls: Charon the ferryman guides souls across the river Styx to the realm of the dead. Souls are judged by Minos, Rhadamanthus and Abacus. They will be sent to one of three realms based on their life: Elysium (paradise), Asphodel Meadows (a neutral place) or Tartarus (a place of punishment.)

Orpheus: A legendary musician travels to the Underworld to retrieve his wife Eurydice - Hades agrees but he must not look back at her until they reach the surface. Upon nearing the exit, Orpheus looks back - causing her to disappear forever.

Celestial Objects & Constellations

CHIRON

The learned centaur who tutored Achilles, Asclepius, Hercules, Jason, and other heroes. Chiron was well versed in medicine, music, prophecy, and hunting, having been raised and educated by Apollo and his wife, Artemis. Chiron was the son of the god Cronus and the sea nymph Philyra.

In modern astrology, Chiron represents our core wounds and how we can overcome them. Chiron is named after a Greek healer, philosopher, and teacher who, ironically, could not heal himself, and is symbolized by a key, demonstrating the importance of unlocking this minor planet's major lessons.

GENERAL: Medicine, music, prophecy, hunting, traumas and vulnerability.

Chiron

Chiron is a centaur; half-human and half-horse. He is known for his intelligence, kindness and nobility as opposed to other centaurs. He is the son of Kronos and Philyra (sea nymph.) He is renowned for his role as teacher of Achilles, Asclepius and Hercules. He has extensive herb and healing knowledge.

Symbols: Staff, scepter, herbs/healing plants, bow and arrow, books/scrolls, horse's body and the constellation Centaurus.

- - -

Immortality: Chiron is immortal unlike most centaurs. However, he suffers a wound that causes him great pain (from Hercules accidentally shooting him with an arrow dipped in the blood of the Hydra) - so much that he relinquishes his immortality in exchange for the release of Prometheus.

Demeter: Upon Persephone's kidnapping, he aids Demeter in her grief by teaching her the healing properties of plants.

Chiron and the Argonauts: Associated with the Argonauts, a group of heroes who accompanied Jason on his quest for the Golden Fleece. He taught the heroes about navigating. music and healing to help them on their quest.

Constellation: Centaurus

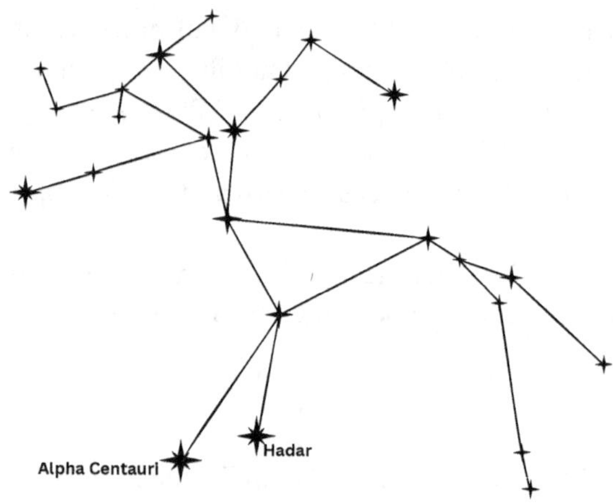

(Shek, Iryna.)

<u>*Can be seen...*</u>

Southern Hemisphere: *High in the sky, year-round. Most visible during Autumn.*

Northern Hemisphere: *Most visible from April to June.*

Notable Facts:

This constellation holds Alpha Centauri, the closest star system to us at 4.37 light years away. It holds three stars - Alpha Centauri A, Alpha Centauri B and Proxima Centauri (the closest star to us.) P.C. also holds a hospitable planet within it's orbit.

Beta Centauri (Hadar) is another bright star located at Centaurus' foot.

LILITH

Across various cultures, Lilith has different stories. In one version, she is seen as the first wife of Adam who would not bow down to him. She was made from the same soil as him and left the Garden of Eden. Then Eve was made from Adam's rib so that women would obey man. She is also recognized as a female demon in others.

Lilith is known as the "dark moon" or the "unseen planet". In astrology placements it can represent untamed energy which may be unaccepted or demonized by others. It could be something one is passionate about that they are encouraged to stifle in order to make others feel at peace.

GENERAL: Intuition, psychic abilities, and hidden knowledge.

Lilith

Lilith is primarily from a Jewish myth. She symbolizes feminine power, independence and rebellion.

Symbols: Owl, serpent, moon, Black Moon Lilith, wings, apple, tree of life and the color red.

- - -

Adam and Eve: Before Eve, there was Lilith. She was made of the same material as Adam. There was only one issue - she refused to submit to Adam and fled the Garden of Eden. God sent angels after her to bring her back but she refused. Because of this, they cursed her to lose 100 of her children a day. God then created Eve from his rib so that she would be submissive.

After leaving Adam she is known to be a succubus - seducing men in their sleep and harming infants. In other lore she is known as the mother of demons.

Black Moon Lilith

Black Moon Lilith is an astrological point, not an actual place, constellation or specified celestial object. It varies from person to person via their astrological birth chart. Specifically, it's the furthest point from the Moon's orbit at any set time, aka the "lunar apogee."

What it represents:

- Repressed desires
- Shadow self
- Feminine power
- Rebellion
- Unconventional sexuality
- Raw, untamed energy

The placement of B.M.L. by sign and house reveals where one may struggle in those areas - notably with power, control or feeling misunderstood. It can also reveal taboo desires or where you should embrace your desires despite potential rejection.

SERPENS

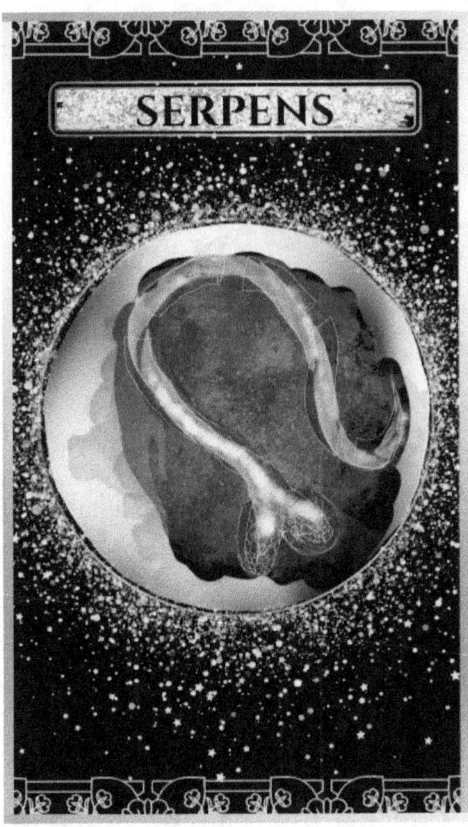

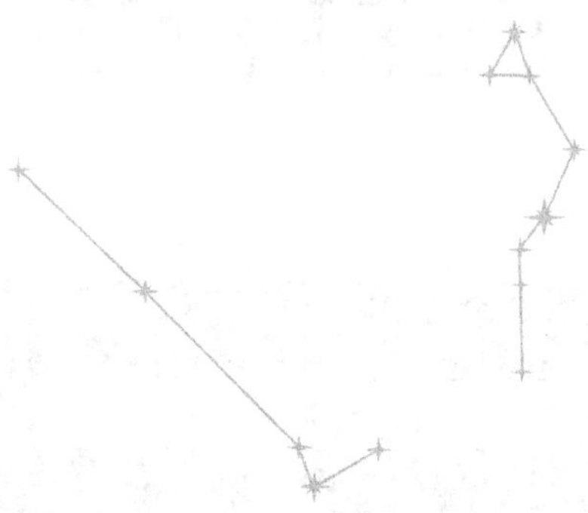

In Greek mythology, Serpens constellation represents a giant snake held by the healer Asclepius, represented by Ophiuchus constellation. Asclepius is usually depicted holding the top half of the snake in his left hand and the tail in his right hand.

GENERAL: Healing, resurrection and medicine.

Constellation: Serpens

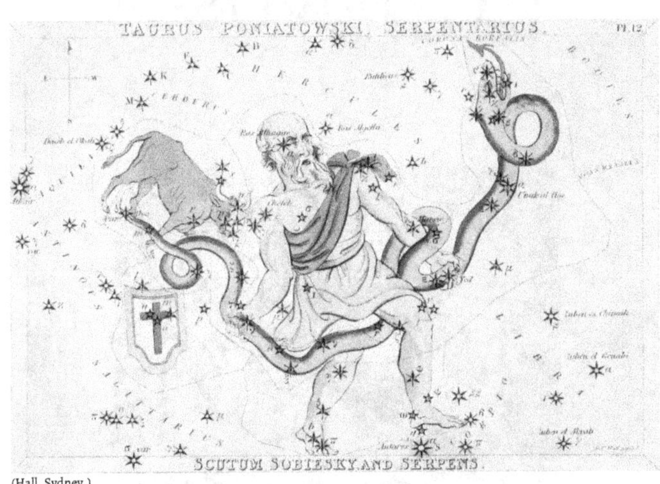

(Hall, Sydney.)

Can be seen...

Southern Hemisphere: *June to August.*

Northern Hemisphere: *April to August.*

Notable Facts:

The Serpens Cloud is a large molecular cloud with regions of
active star formation in Serpens Cauda.

Alpha Serpentis, or Unukalhai, is the brightest star within
the constellation (which is split into two, as
Asclepius/Ophiuchus holds it.)

CYGNUS

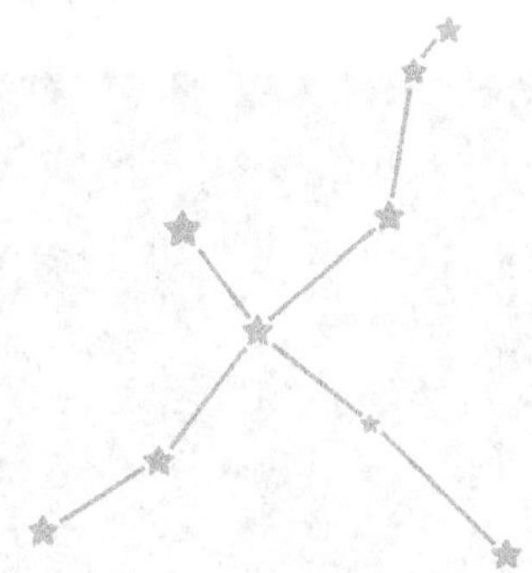

In Greek mythology, Cygnus represents a swan. There are multiple myths surrounding this constellation. One of them states Phaeton, Helios' mortal son and Cygnus were close friends. When racing too close to the sky, their chariots burned up and fell to Earth. Cygnus found Phaeton in the Eridanus river. Unable to recover the body, he made a pact with Zeus: to give him the body of a swan as well as a swan's lifespan so that he may dive into the river and retrieve his friend's body for a proper burial. Because of this, Phaeton was able to go to the afterlife. Zeus was pleased by this and put his constellation in the sky.

GENERAL: Selflessness, sacrifice and care for others.

Constellation: Cygnus

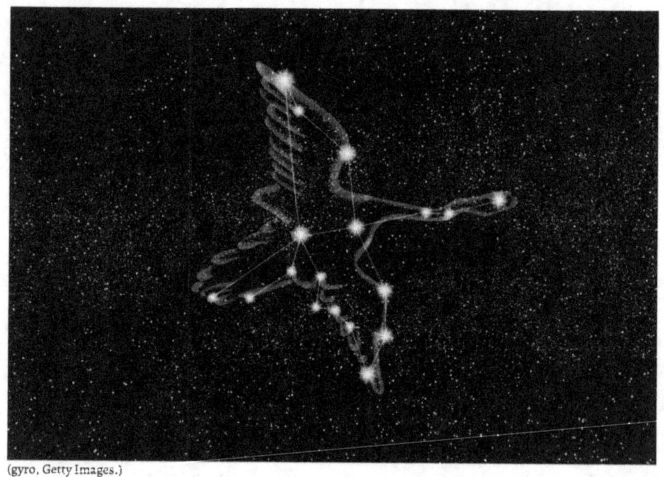

(gyro, Getty Images.)

<u>Can be seen...</u>

Southern Hemisphere: *June to September.*

Northern Hemisphere: *Late June to early September (more visible within this region.)*

Notable Facts:

The brightest star is Deneb (Alpha Cygni), a blue-white super giant located 2,600 light years away. Albireo (Beta Cygni) is a binary star system with a gold and blue star.

Cygnus is home to the Veil Nebula (NGC 6960, NGC 6992), a supernova popular amongst astrophotographers.

ORION

Orion was said to be a giant and handsome huntsman. There are multiple myths surrounding him. In the oldest version, he was the son of Poseidon and Euryale, daughter of King Minos of Crete. Orion could walk on water, allowing him to go to Chios the island. There he made sexual advances towards Merope drunkenly, a daughter of the King Oenopion who had him blinded and removed from the island. Helios restored his eyesight. Later, he would hunt with Artemis and declared he would kill every animal in the world. Gaia was displeased and sent Scorpius the giant scorpion to kill him. Zeus placed him and Scorpius in the sky.

GENERAL: Boldness, audacity and strong-willed.

Constellation: Orion

(manpuku7, Getty Images.)

Can be seen...

Southern Hemisphere: *Summer.*

Northern Hemisphere: *December - March.*

Notable Facts:

Betelgeuse, a red supergiant, is located on his right shoulder. The brightest star in this constellation is a blue supergiant, Rigel, located in his left foot.

Stargazing tip:

Orion's belt is an easy way to spot this formation - 3 stars in a line; Alnitak, Alnilam, and Mintaka.

CEPHEUS

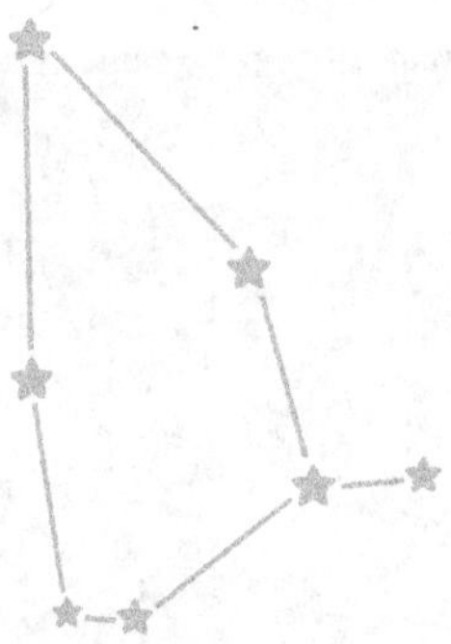

King Cepheus of Aethiopia was forced to sacrifice his daughter Andromeda to the sea monster Cetus by Poseidon after his wife Queen Cassiopeia proclaimed she was more beautiful than the Nereids.

Cepheus went to an oracle on how to go about this situation. The woman said he would have to sacrifice his daughter to Cetus. He chained his daughter to a rock for the monster.

GENERAL: Desperation, haste and quick decisions.

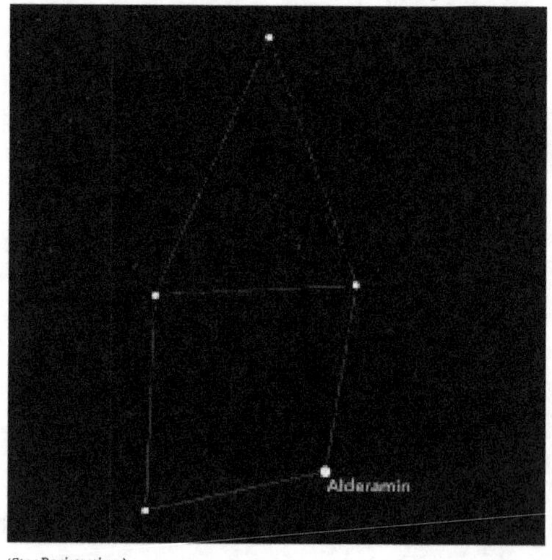

(Star Registration.)

Can be seen...

Southern Hemisphere: *Rarely, and one must be located closer to the Northern Hemisphere to view it during September - December.*

Northern Hemisphere: *Annually.*

Notable Facts:

This constellation is in the shape of a house/crown. It is circumpolar, meaning that it never sets and is visible year-round in the Northern Hemisphere.

Within this constellation are cepheid variable stars, meaning that they pulsate with different amounts of light.

CASSIOPEIA

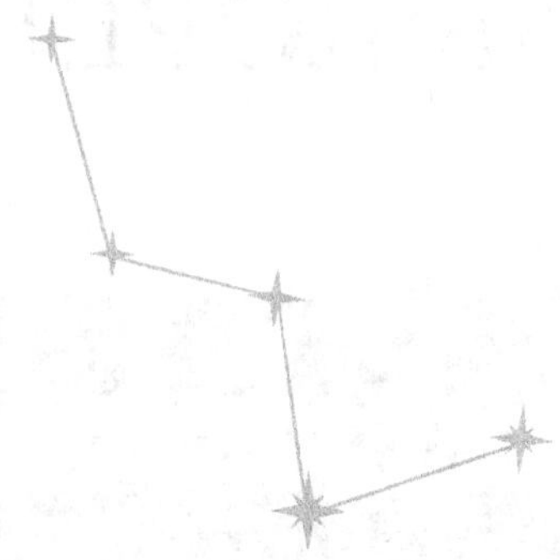

Queen Cassiopeia proclaimed she was more beautiful than the Nereids. Because of this, her husband King Cepheus and her after consulting an oracle would tie their daughter to a rock after having a sea monster Cetus be sent by Poseidon to destroy everything.

When things did not go as planned, Poseidon was not happy that she went unpunished. He tied her to a chair in the heavens, so that she would revolve upside down half of the time. The constellation represents the torture chair she sits in.

GENERAL: Vanity, ego, lack of care for others and consequences.

Constellation: Cassiopeia

(Planet Guide.)

Can be seen...

Southern Hemisphere: *Rarely, and one must be located closer to the Northern Hemisphere to view it during September - February.*

Northern Hemisphere: *Annually.*

Notable Facts:

Cassiopeia holds 5 stars: Schedar, Caph, Gamma Cassiopeiae, Ruchbah, and Segin. Like Cepheus, it is also circumpolar, near Polaris, which will be discussed later on.

Within this region resides the Pacman Nebula - named after its resemblance to the video game character.

ANDROMEDA

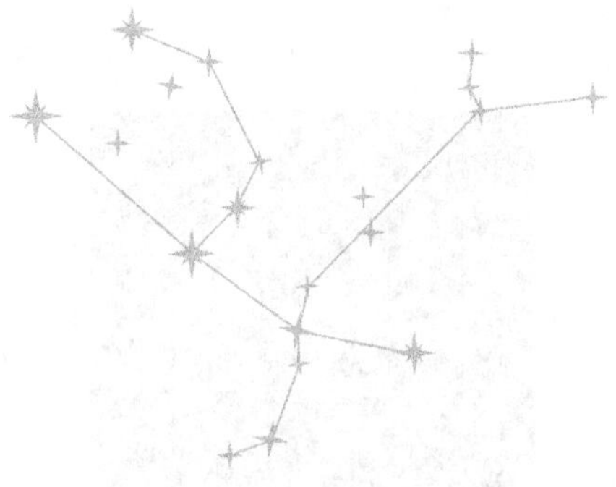

Andromeda was naked, chained to a rock and sacrificed to the sea monster Cetus due to her mother's vanity. Perseus, whilst riding Pegasus, sees her and asks Cepheus to marry her if he can save her. He does, later taking her to be his wife in Greece as Queen.

She was previously engaged to her Uncle, Phineus. At the wedding, a fight breaks out and Persues uses Medusa's head to turn Phineus and his men into stone.

GENERAL: Innocence, beauty, engagement and marraige.

Constellation: Andromeda

(Star Name Registry.)

Can be seen...

Southern Hemisphere: *September - February.*

Northern Hemisphere: *August - December.*

Notable Facts:

This constellation also holds the Andromeda Galaxy - the closest fully-fledged galaxy to our own at 2.5 million light years away. It is one of the furthest objects one can see with their naked eye.

Andromeda is the 19th largest constellation out of the modern 88.

LUPUS

The Lupus constellation in modern day represents a wolf. In the past, it was not considered that, but a wild and unknown animal. The Greeks called it Therion and the Romans called it Bestia ("beast"). Both cultures originally believed this animal was given as a sacrifice to an altar. The constellation can sometimes be seen as held by Centaurus.

GENERAL: Ceremony, altar, sacrificial ceremony.

Constellation: Lupus

(Flickr.)

Can be seen...

Southern Hemisphere: *May - August*

Northern Hemisphere: *April - June*

Notable Facts:

Close to Centaurus, it is seen as a wolf being speared by him. It used to be associated with the Centaurus constellation as one as well. This constellation was not a part of Greek mythology.

Lupus is a dim constellation and can be hard to spot due to light pollution. It's located near the Milky Way plane. Lupus wasn't categorized as a wolf until the Renaissance.

POLARIS

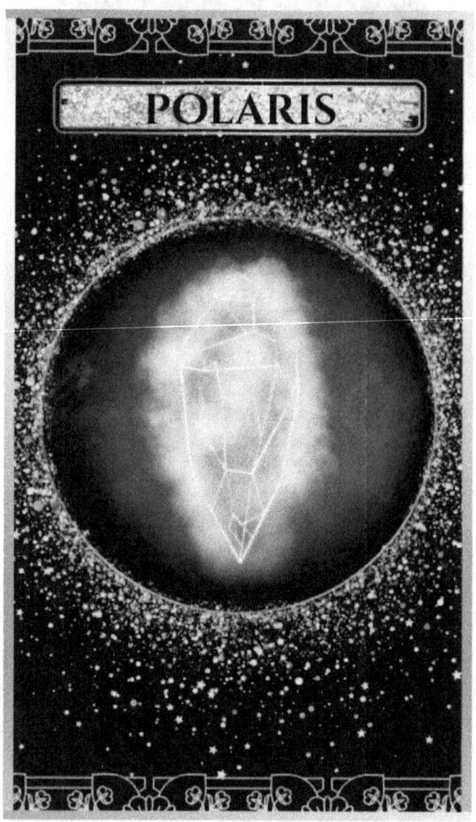

Polaris is a star located in the Ursa Minor constellation. It is also known as the North Star or the Pole Star. It is the brightest star within the constellation and can be seen with a naked eye. It is only one degree away from the North Celestial Pole.

GENERAL: Guidance, starting point

Star: Polaris

(EarthSky)

<u>*Can be seen...*</u>

Southern Hemisphere: *September - February.*

Northern Hemisphere: *August - December.*

Notable Facts:

Located close to Earth's Northern Pole, making it a important navigational point. Despite it's notoriety it's only the 50th brightest star in the night sky.

This star can be found in the Little Dipper. It will not always be the North star due to Earth's axial precession. In 13,000 years Vega (within Lyra) will be the true North star.

DELPHINUS

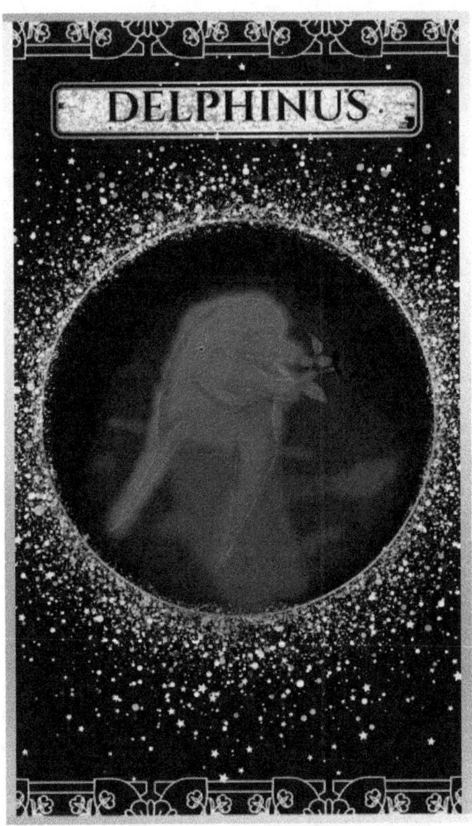

Delphinus was Poseidon's dolphin messenger. The sea god tried to court Amphitrite, to which she refused his advances. Poseidon sent out his messengers, including Delphinus, to find her. The dolphin soothed her and brought her back to him - and later the two would marry. Poseidon put him in the sky to honor him.

GENERAL: Assistance in love and love letters.

Constellation: Delphinus

(EarthSky).

Can be seen...

Northern Hemisphere: *August - October.*

Notable Facts:

This constellation can be found in between Pegasus and Aquila. Deep-sky objects within this star link are: NGC 6891 (a planetary nebula with a bright, central star), NGC 6905 (the Blue Flash Nebula) and the Delphins Galaxy group.

CORVUS

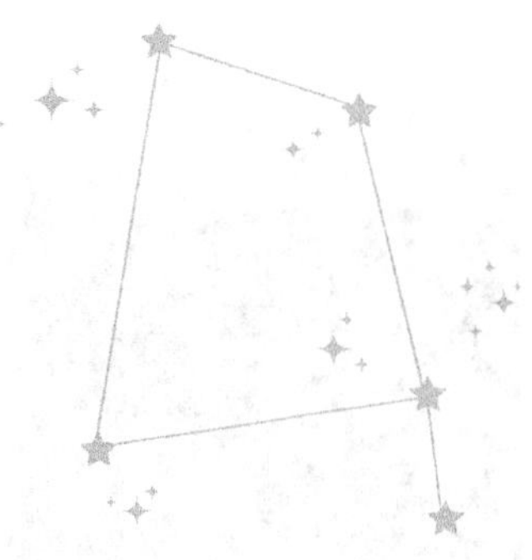

Corvus was a crow, Apollo's sacred bird. He was given a chalice and sent to collect the waters of life. Instead of immediately doing so, he waited around a fig tree for the fruit to ripen. When he finally returned to Apollo, he made up excuses stating that a watersnake delayed him.

Apollo did not believe this and cursed the bird with eternal thirst and threw him into the sky.

GENERAL: Mission, deception, delay and excuses.

Constellation: Corvus

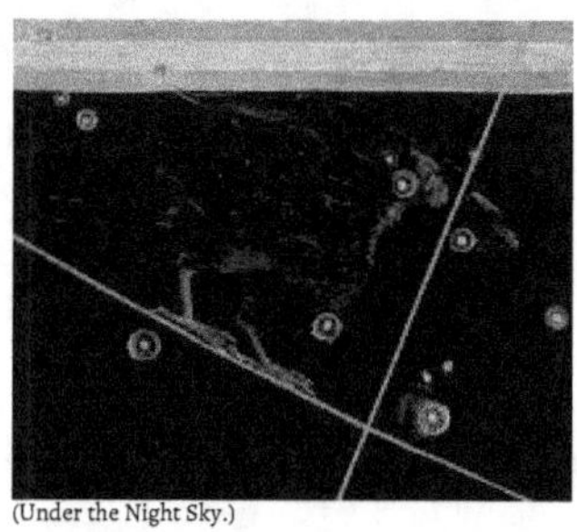

(Under the Night Sky.)

<u>*Can be seen...*</u>

Northern Hemisphere: *March - May.*

Southern Hemisphere: *May.*

Notable Facts:

A quadrilateral star formation located in the Southern Hemisphere. It is a part of a trio with the constellations Hydra and Crater.

Gienah (Gamma Corvi) is the brightest star and means "wing" in Arabic. Algorab (Delta Corvi) is slightly dimmer and recognized as part of the "raven" shape.

NGC 4038 & 4039 are two colliding galaxies found here.

HYDRA

Hydra was blamed by Corvus for his tardiness. Hydra is also associated with the second labor of Hercules. It was a multi-headed monster of Typhon and Echidna. It's mother was half-woman, half-serpent. The dragon Ladon which guarded the garden of Hesperides was Hydra's brother.

In mythology, Hydra had 9 heads - one of which was immortal. It lived near Lerna, a town it ravaged. Hercules smoked it out of its cave to fight it, smashing its heads - but new heads would sprout. His charioteer Iolaus helped by burning the neck stubs so more heads wouldn't grow. Hercules then cut off the immortal head and buried it under rocks.

GENERAL: Scapegoat, new issues arise.

Constellation: Hydra

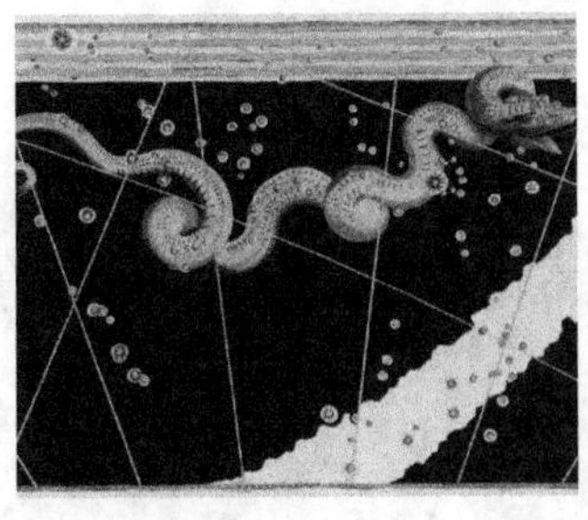

(Under the Night Sky)

Can be seen...

Northern Hemisphere: *January - May.*

Southern Hemisphere: *February - March, Autumn.*

Notable Facts:

Hydra is the largest constellation in the sky and covers 1,303 square degrees. It doesn't contain many bright stars, causing it to be harder to spot.

This formation lies between Cancer and Libra with its tail reaching Centaurus. The Hydra Cluster features over 100 galaxies at 150 million light-years away.

PERSEUS

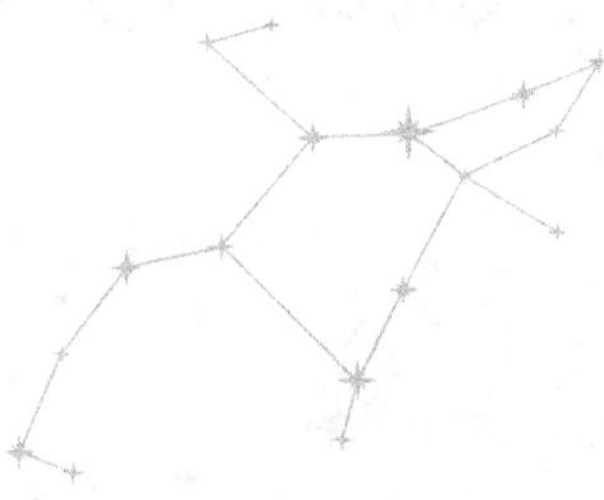

Perseus, son of Zeus and Danaë, husband of Andromeda, King of Greece. He slayed the gorgon Medusa and saved Queen Andromeda from being eaten by Cetus the sea monster. He fulfilled the prophecy of killing his grandfather - doing so by accident via discus.

GENERAL: Heroism, prophecy, assistance and adventure.

Constellation: Perseus

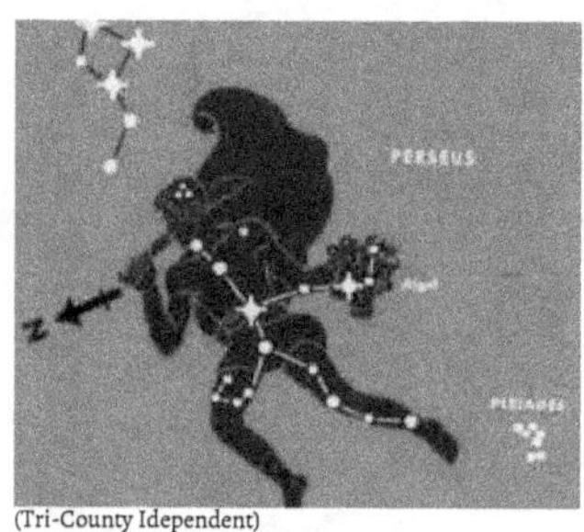

(Tri-County Idependent)

Can be seen...

Northern Hemisphere: *August - March.*

Notable Facts:

This constellation can be found between Andromeda and Auriga. Perseus holds the Perseids meteor shower, one of the most visual and recurrent meteor showers. It occurs annually in mid-August and is caused by debris via the comet Swift-Tuttle.

You can see the Double Cluster, a pair of open star clusters, with your naked eye here. Perseus belongs to the myth star family including Andromeda, Cepheus, Cassiopeia and Cetus.

PEGASUS

Pegasus was a winged horse with magical powers. It's hooves dug out a river named Hippocrene which gave those who drank from it the talent of poetry. It was birthed from Medusa's stomach after she was slain, it's father being Poseidon.

It would become the horse for Bellerophon, who was to slay the Chimera. Afterwards it took him to Olympus, but Zeus sent a thunderbolt at them.

GENERAL: Blessing, partnership and poetry.

Constellation: Pegasus

(Under the Night Sky)

Can be seen...

Northern Hemisphere: *September - January.*

Notable Facts:

The Great Square, an asterism formed by Markab, Scheat, Algenib and Apheratz ((stars) Alpheratz is a part of Andromeda), is easy to spot in the sky and is used to find other constellations by stargazers.

It is the 7th largest constellation. Enif is the largest star - a orange supergiant - located 670 light-years from Earth.

ERIDANUS

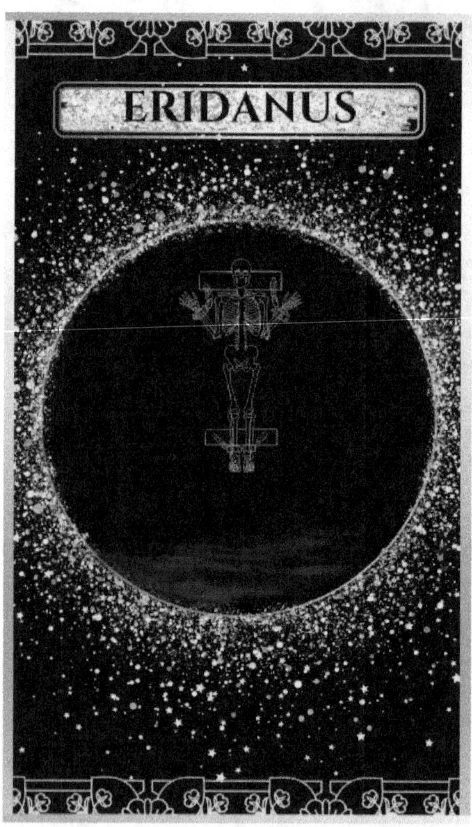

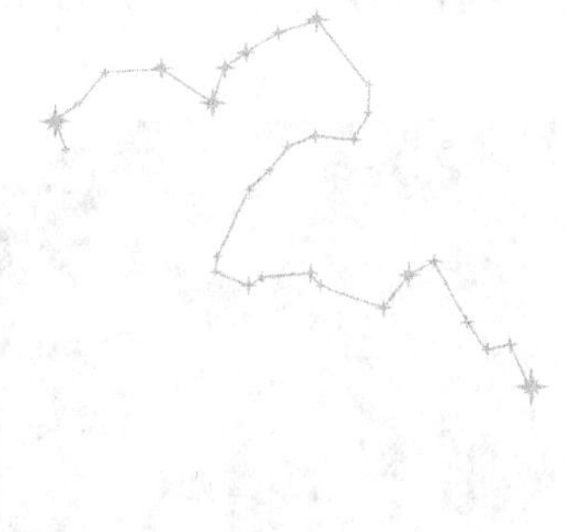

Eridanus was known as a river, sometimes associated with the Nile and the Po. It was said to go around the world. It is the sixth largest modern constellation. It is also known as Hades' river.

In relation to the card Cygnus, Phaeton's body was found at the bottom of the Eridanus River. Much of the river was evaporated by Phaeton being so hot from falling from the sky.

GENERAL: Water and travel.

Constellation: Eridanus

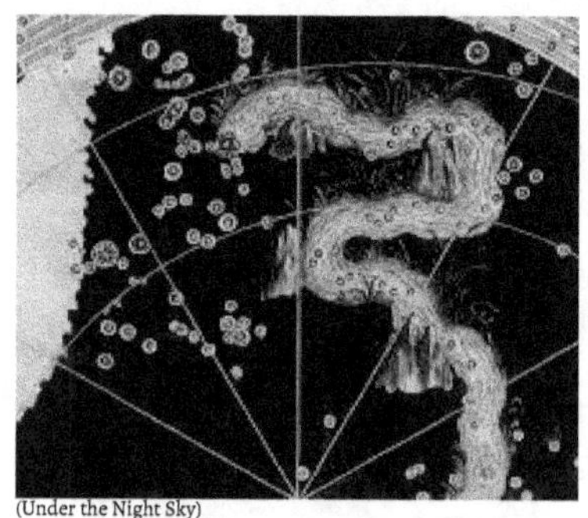

(Under the Night Sky)

<u>**Can be seen...**</u>

Southern Hemisphere: *November - February.*

Southern Hemisphere: *Winter.*

Notable Facts:

This constellation stretches from Orion (north) to the bright star Achernar in the south. Archernar means "the river's end" in Arabic and is a blue giant. The star Cursa (Beta Eridani) marks the beginning of the river.

The Eridanus Supervoid is a region of space significantly empty of galaxies and matter, puzzling to astronomers due to it being larger than expected via current cosmological theories.

EQUULEUS

Equuleus means "little horse" or "foal". It is associated with multiple myths - the main one being of Hippe. She was the daughter of Chiron and
Chariclo.

Equuleus was seduced by Aeolus and became pregnant with his child. She was ashamed and hid in the mountains until she gave birth to her child Melanippe.

Chiron would come looking for his daughter, to which she asked the gods to not be found. In response, she was turned into a foal and placed amongst the stars.

GENERAL: Shame, secrecy and hiding.

Constellation: Equuleus

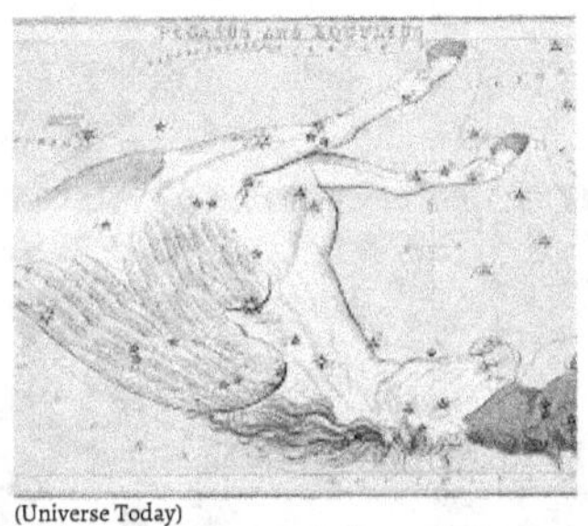

(Universe Today)

Can be seen...

Northern Hemisphere: *Fall.*

Notable Facts:

Equuleus is the second smallest constellation. It's next to Pegasus and Delphinus.

It holds NGC 7013, a spiral galaxy located 100 million light-years away. There are no notable star clusters.

OCTANS

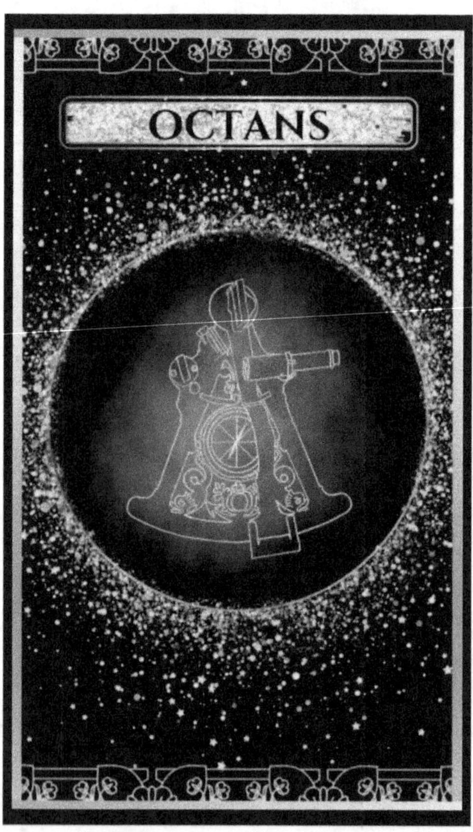

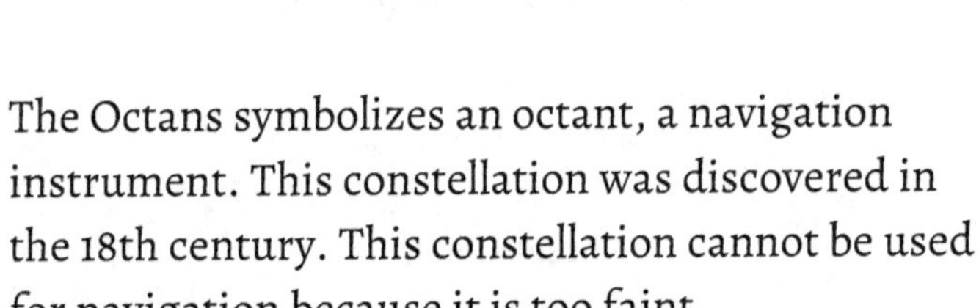

The Octans symbolizes an octant, a navigation instrument. This constellation was discovered in the 18th century. This constellation cannot be used for navigation because it is too faint.

The southern celestial pole is located within the boundaries of Octans. It is most visible during the month of October. This constellation does not have any mythology tied to it.

GENERAL: Unclear navigation.

Constellation: Octans

(Star Registation)

Can be seen...

Southern Hemisphere: *Annually.*

Notable Facts:

Alpha Octantis, aka Hydrus, is the brightest star in this constellation. Octans wields Polaris Australis - the Southern Pole Star.

This constellation serves mostly as a tool for mapping and navigation - particularly sailors and explorers within the Southern Hemisphere. It is used to determine latitude and orientation.

PYXIS

Pyxis represents a mariner's compass.

This constellation does not have any mythology tied to it. Pyxis' has three main stars (others which are not as visible). This constellation is best witnessed during the month of March.

GENERAL: Navigation, sea and ships.

Constellation: Pyxis

(Go-Astronomy)

Can be seen...

Southern Hemisphere: *Autumn and Winter.*

Northern Hemisphere: *February - March.*

Notable Facts:

Pyxis is near the constellation Argo Navis, which represents the ship Argo from Greek mythology. It is also between Vela (the Sails) and Carina (the Keel.)

One of the most notable objects in this region is NGC 2750, a barred spiral galaxy located about 70 million light-years away.

CETUS

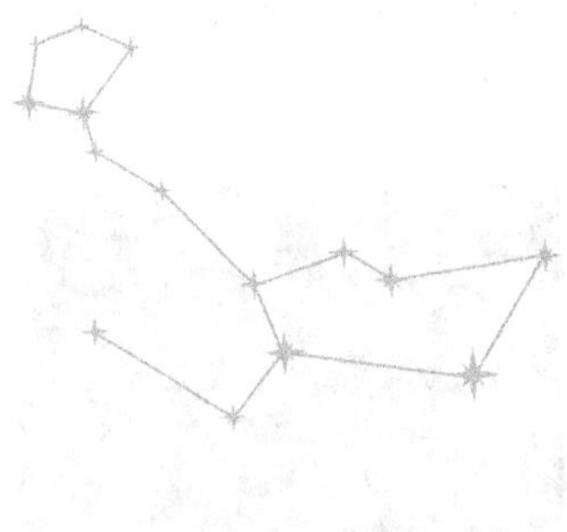

Cetus is any huge sea monster, usually represented as a whale. In mythology, Poseidon sent him after Ethiopia after the Queen boasted that her and her daughter were more beautiful than the Nereids. The royalty were told to sacrifice their daughter to Cetus. Perseus would save their daughter (Andromeda) by slaying the monster.

Cetus is not a part of the traditional zodiac signs, although some consider it the 14th sign as the planets briefly pass through it.

GENERAL: Wrath, gates to the underworld and water magic.

Constellation: Cetus

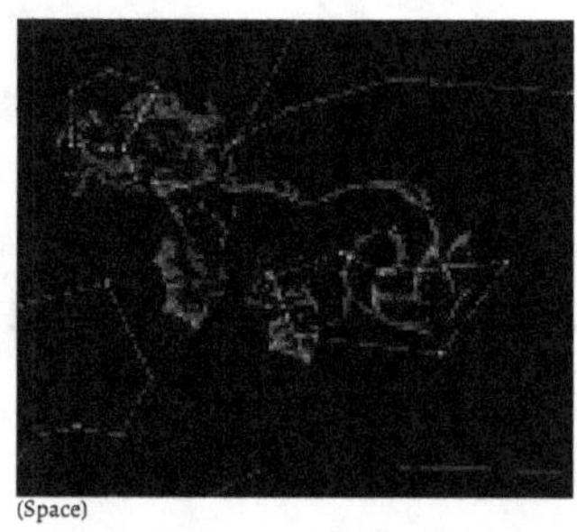

(Space)

Can be seen...

Southern Hemisphere: *September - December.*

Northern Hemisphere: *October - March.*

Notable Facts:

This constellation is near Pisces, Aries and Aquarius. Deneb Kaitos (Beta Ceti)/ "the tail of the whale" is a giant star 2.5 times larger than the Sun with a magnitude of 4.54.

NGC 246, or "Cetus Nebula", is a bright planetary nebula found within this region. The Cetus Supernova was observed in 1572 by astronomer Tycho Brahe. That changed our understanding of celestial objects and led to developments in astronomy.

URSA MAJOR

Ursa Major is the Great Bear. In Greek mythology, Callisto was a follower of Artemis and a nymph. Her name meant "the most beautiful" and was sworn to be a virgin throughout her life. She would later hold a child with Zeus. When Artemis found out, she was so angry that she turned her into a bear, to be killed by hunters. She would deliver her son in bear form, and he would be King of Arcadia. In the future, the son would be close to killing a bear (his mother) before Zeus turned them into constellations - Ursa Major and Minor.

The Big Dipper constellation is a part of Ursa Major.

GENERAL: Motherhood, separation from child, infidelity and broken promises.

Constellation: Ursa Major

(Star Registration)

Can be seen...

Southern Hemisphere: *March - August.*

Northern Hemisphere: *Annually.*

Notable Facts:

Ursa Major (accompanied by Ursa Minor/the Little Dipper) is the third largest constellation. This constellation has been used for navigational purposes for centuries. The position of the dipper relative to Polaris allows travelers to find true north.

Messier 81 (M81) and Messier 82 (M82) are two interacting galaxies, also popular targets for amateur astronomers. Messier 97 (M97) is the Owl Nebula (planetary.)

CANIS MAJOR

Canis Major, in mythology, was the fastest dog. He was so fast that no prey could outrun him. He was used to hunt down the Teumessian Fox, a beast determined to never be caught. The chase was endless until Zeus finally turned them both into stone.

Canis Major follows Orion in pursuit of a hare, Lepus along with Canis Minor. The constellation is built around the star Sirius, the brightest star in the sky.

GENERAL: Quick, speed, endless chase and agility.

Constellation: Canis Major

(Under the Night Sky)

Can be seen...

Southern Hemisphere: *December - April.*

Northern Hemisphere: *December - March.*

Notable Facts:

1 of the 2 hunting dogs of Orion. Canis Major is home to Sirius, the brightest star in the night sky. It is also known as the Dog Star.

Close to this constellation are Lepus, Puppis and Monoceros. In Egyptian culture, the rising of Sirius marked the annual flooding of the Nile, also seen as a time for rebirth.

DRACO

Draco represents Ladon, a serpent-like dragon who guarded the golden apples in the Garden of Hesperides. It was Hercules' 11th labor to kill him, which he did with a bow and arrow, and take the apples. In a different version, Hercules temporarily exchanges places with Atlas holding the globe so he may retrieve them - and Ladon is not slain.

Draco is one of the largest constellations. It revolves around the celestial North Pole. The tail ends near Ursa Major and Minor.

GENERAL: Protection and guardianship.

Constellation: Draco

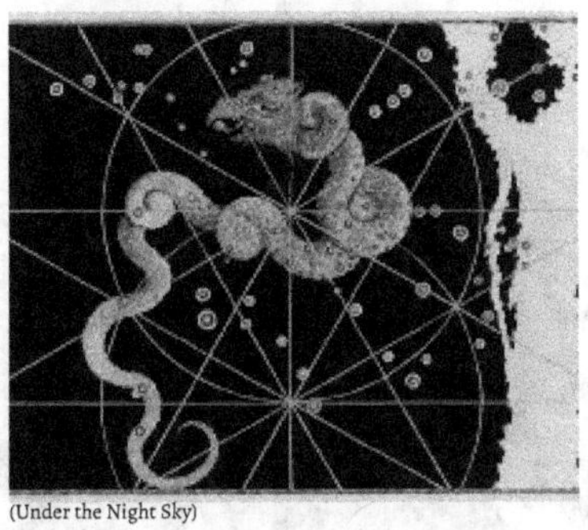

(Under the Night Sky)

Can be seen...

Northern Hemisphere: *Annually.*

Notable Facts:

8/88th in ranked size. This constellation is between Ursa Major and Ursa Minor. Thuban (Alpha Draconis), is the brightest star within it and was once the North Star around 2700 BCE.

Messier 1 (M1) is the Crab Nebula (supernova remnant.) NGC 5866, Tetrahedron Galaxy/Spindle Galaxy, is a spiral galaxy above Draco. Draco has severable asterisms including Draco's Circlet in the dragon's head.

CRUX

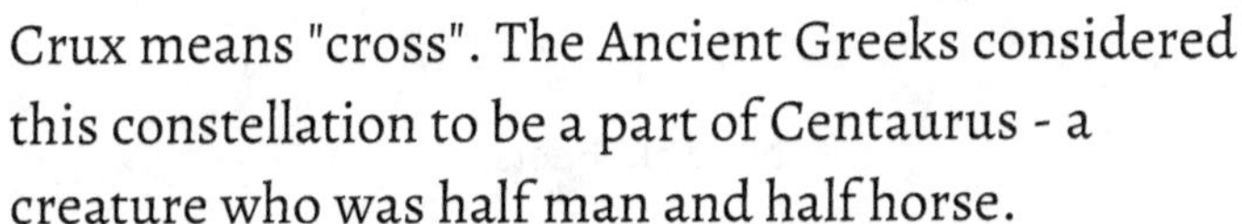

Crux means "cross". The Ancient Greeks considered this constellation to be a part of Centaurus - a creature who was half man and half horse.

Crux is the smallest of all constellations and is located in the Southern sky. Augustin Royer separated Centaurus to create Crux in 1613.

GENERAL: Religion, transformation, sacrifice and imposition.

Constellation: Crux

(Star Registration)

<u>*Can be seen...*</u>

Southern Hemisphere: *April - August.*

Notable Facts:

Acrux (Alpha Crucis) is a binary star system and one of the brightest in the Southern Hemisphere. Mimosa (Beta Crucis) is a blue giant and Gacrux (Gamma Crucis) is a red giant.

Crux is used for navigational purposes in the south. By extending the long axis of the cross by 4.5 times in length, one can find the approximate position of the South Celestial Pole.

CARINA

The constellation Carina holds the second brightest star in the night sky. This star, Canopus, aids in modern space exploration. In Latin, Carina refers to the keel of a ship. It used to be a part of a larger constellation representing the ship Argo.

Jason and the Argonauts sailed on Argo in search of the Golden Fleece of Chrysomallos, the winged ram. The ship was built by the help of the gods. Athena made it so it could handle the longest voyage every made. With the help of Medea, they stole it from King Aeëtes.

GENERAL: Authority, kingship, inner world and space exploration.

Constellation: Carina

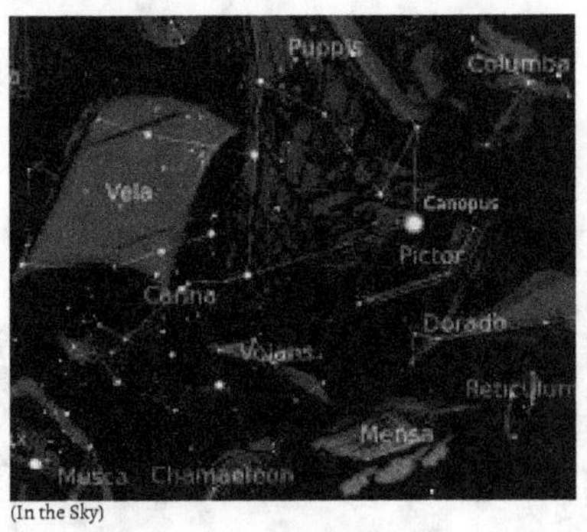

(In the Sky)

Can be seen...

Southern Hemisphere: *December - February.*

Notable Facts:

Carina is a part of Argo Navis, which is divided into three parts: Carina (the keel), Vela (the sails) and Puppis (the stern.)

Carina has a wealth of stars and stellar phenomena. It's studied to learn about star formation and the life cycles of massive stars.

To view active star formation, check out the Carina Nebula or NGC 3372. It also contains some of the most massive known stars.

TITAN

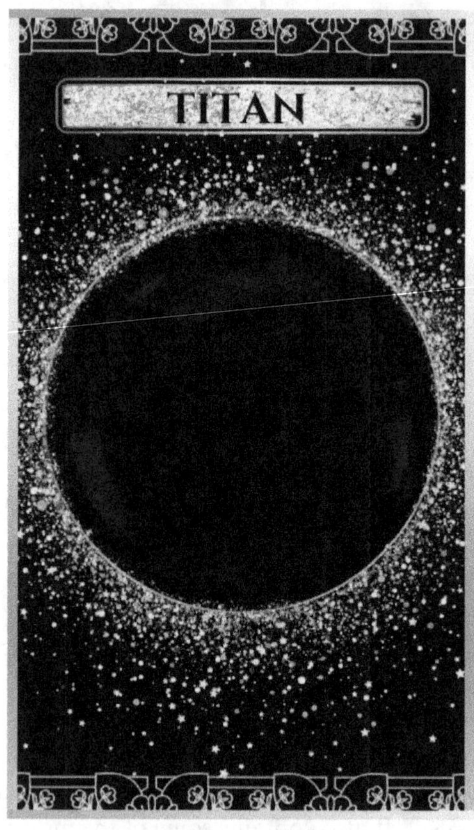

The Titans were the predecessor Gods to those like Zeus and Hades (pre-Olympian). They were banished beneath the Earth into Tartarus.

In astrology Titan rules the past in conjunction with Saturn. Titan can represent the national, racial, and economic situations and issues we were born among, whether privileged or disadvantaged. It can represent "karma" and its resolution. The past can hold us back, or it can be a source of advice and assets, and act as "shoulders for us to stand upon".

Titan is Saturn's largest moon. It is the only moon with a thick atmosphere and pools of water like Earth. Titan is bigger than Mercury.

GENERAL: Past, entrapment, power and upbringing.

Moon: Titan

(Space)

Notable Facts:

Titan is Saturn's largest moon, and the second-largest moon within our solar system. It's comprised mostly of nitrogen with small amounts of methane/other gases.

Titan is the only other celestial body other than Earth to be known to have stable bodies of liquid on the surface. It has large lakes and rivers of liquid methane and ethane within (but not limited to) it's polar regions.

Titan is believed to have cryovolcanoes (cold volcanoes.) It's subsurface ocean is thought to possible harbor microbial life.

Titan has synchronous rotation with Saturn, meaning it always shows the same face to the planet. It orbits every 15.9 days, which is also it's rotational period.

NEBULA

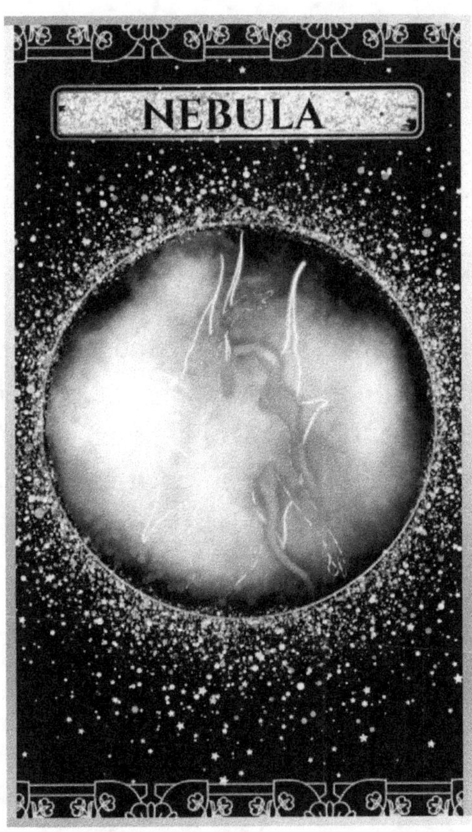

A nebula is a giant cloud of dust and gas in space. Some nebulae (more than one nebula) come from the gas and dust thrown out by the explosion of a dying star, such as a supernova. Other nebulae are regions where new stars are beginning to form.

Nebulae are **symbolic of reincarnation**. They are continually rebirthing the cosmos. Nebulas are crucibles of creation. They are the birth of new star systems and the beginning of light.

GENERAL: Hope, reincarnation, new beginnings, building blocks.

Nebula

(Webb Space Telescope)

Notable Facts:

A nebula is a giant cloud of gas and dust. It can be a region where new stars are born, the remnants of a dead or dying star, or simply large clouds. There are different types of nebulae:

Emission Nebulae glow brightly due to ionization of gas via nearby hot stars. They're often red because of hydrogen.

Reflection Nebulae don't emit their own light but reflect light from nearby stars. They often appear blue.

Dark Nebulae are dense regions of gas/dust that block light from objects behind them. They appear as dark silhouettes.

Planetary Nebulae is when a star exhausts its nuclear fuel and expels its outer layers - leaving behind a hot core that ionizes expelled gas.

COMET

Comets are **frozen leftovers from the formation of the solar system composed of dust, rock, and ices.** They range from a few miles to tens of miles wide, but as they orbit closer to the Sun, they heat up and spew gases and dust into a glowing head that can be larger than a planet.

Comets have been branded with such titles as "the Harbinger of Doom" and "the Menace of the Universe." They have been regarded both as omens of disaster and messengers of the gods.

GENERAL: Disaster, messages from afar, new thoughts and changes.

Comet

Notable Facts:

Comets are icy bodies that when they come close to a star, heat up and release gases, creating a glowing coma that is usually accompanied by a tail. This is from when ice sublimates directly into gas.

Comets are described as "dirty snowballs" because they're composed of ice, dust and rocky material. Comes usually have two tails: a gas (ion) tail and a dust tail. The gas tail points directly away from a star (solar wind) and the dust tail follows the comet's orbit with a curve.

Some comet orbits take hundreds or thousands of years to circuit. Short-period comets take less than 200 years to complete. Long-period comets take over 200 years.

BLACK HOLE

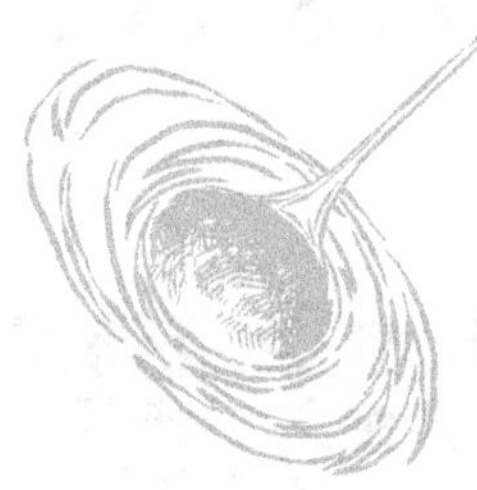

A black hole is a region of spacetime where gravity is so strong that nothing, including light or other electromagnetic waves, has enough energy to escape it. The theory of general relativity predicts that a sufficiently compact mass can deform spacetime to form a black hole.

The first black hole was discovered in 1964, Cygnus X-1, meaning that the Greeks were unaware of them. The idea of the black hole first came about in 1783 by John Michell, and English country parson.

GENERAL: Energy vampire, stealth, dark energy, magnetism, narcissism and encapture.

Black Hole

(Space)

Notable Facts:

Black holes are regions in space where gravity is so strong that nothing can escape from it, not even light. Typically, they are formed when massive stars exhaust their nuclear fuel and collapse under their own gravity. This can occur during a supernova.

The boundary surrounding it is called the event horizon. Once an object crosses it, it cannot escape the gravitational pull. This can be called "the point of no return."

At the center is a point called the singularity - where matter is compressed into an infinitely dense point, breaking the laws of physics as we know them.

There are stellar (stars after supernovae), supermassive (found at centers of galaxies), intermediate (existence not yet proven) and primordial black holes (hypothetical black holes that may have formed during early density fluctuations in the Universe.)

ARIES

Aries rescued Phrixus and took him to Colchis where he sacrificed the ram to appease the Gods. Later, Phrixus, in the face of death was saved by a golden ram with wings which flew him to safety.

The Zodiac sign Aries is ruled by Mars. People born underneath this star sign are characterized as passionate, motivated, and confident leader who builds community with their cheerful disposition and relentless determination. Uncomplicated and direct in their approach, they often get frustrated by exhaustive details and unnecessary nuances.

ELEMENT: Fire

BODY: Head and brain.

GENERAL: Impulsivity, motivation and helping others.

Constellation: Aries

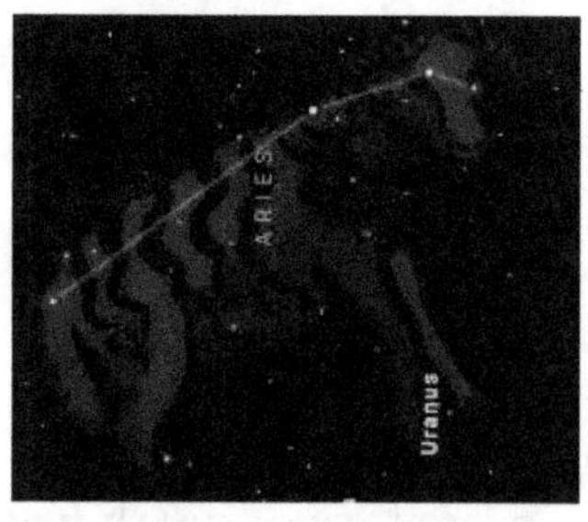

(StarLust)

Can be seen...

Southern Hemisphere: *March - May.*

Northern Hemisphere: *September - December.*

Notable Facts:

1/12 zodiac constellations located in the northern celestial hemisphere. It is next to Pisces (west) and Taurus (east.) In Greek mythology, after being sacrificed its fleece became the coveted Golden Fleece.

The three main stars of Aries form an asterism known as the "Triangle of Aries." It is also associated with the spring equinox.

TAURUS

The constellation Taurus commemorates the god Zeus. Zeus changed himself into a beautiful white Bull to win the affections of the Phoenician Princess Europa.

The Zodiac sign Taurus is ruled by Venus. People born underneath this star sign are characterized as being hard-headed, down-to-earth, tenacious, reliable, loyal, and sensual. Taurus rules the houses of Self-Worth and Income.

ELEMENT: Earth

BODY: Throat, neck, vocal chords, tonsils and thyroid.

GENERAL: Self-importance, decisions, leadership and tenacity.

Constellation: Taurus

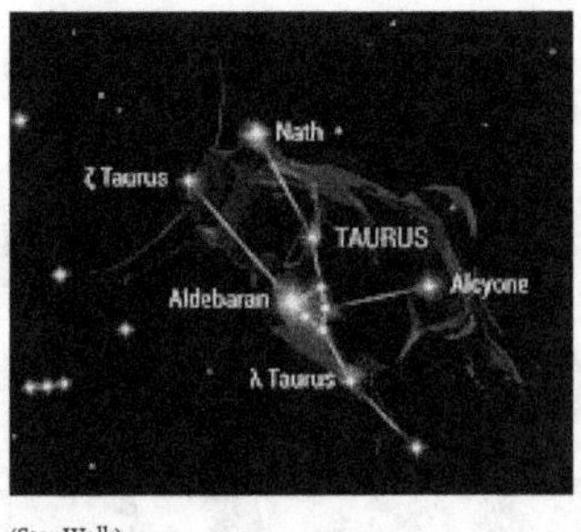

(Star Walk)

Can be seen...

Southern Hemisphere: *December - February.*

Northern Hemisphere: *November - March.*

Notable Facts:

Orion borders this constellation to the south. It's brightest star is Aldebaran (Alpha Tauri), a red giant star and is referred to as the "Eye of the Bull."

Two famous star clusters are here: The Pleiades and the Hyades (forms the V-shaped pattern of the bull's head.)

GEMINI

The constellation of Gemini represents Castor and Polydeuces. They were identical twins born to Leda, Queen of Sparta, by two different fathers. Castor was said to be the son of Leda's husband, King Tyndareus and thus mortal. Polydeuces was said to be the immortal son of Zeus.

The Zodiac sign Gemini is ruled by Mercury. People born underneath this star sign are characterized being communicative, funny, outgoing, inquisitive, impulsive and curious. Gemini is constantly juggling a variety of passions, hobbies, careers, and friend groups.

ELEMENT: Air

BODY: Shoulders, lungs, arms, hands, and fingers.

GENERAL: Curiousity, fun, flexibility and witty.

Constellation: Gemini

(Cosmo Nova)

Can be seen...

Southern Hemisphere: *December - March.*

Northern Hemisphere: *December - February.*

Notable Facts:

Gemini is Latin for "the twins." The brightest star in Gemini is Pollux (Beta Geminorum), a giant star with a magnitude of 1.14. Castor (Alpha Geminorum) has a complex multiple star system comprised of six stars with a magnitude of 1.58.

The Geminids, a prominent meteor shower, occurs annually in December. It also has Thors Helmet Nebula, NGC 2359.

CANCER

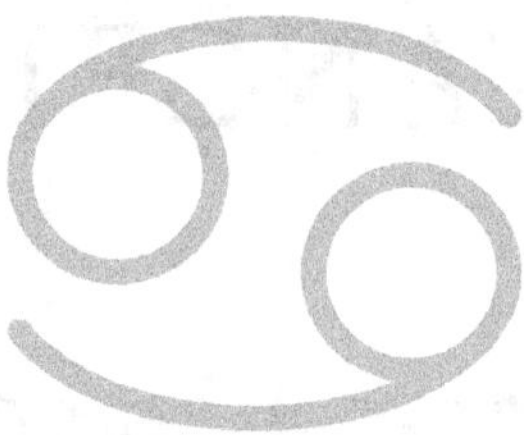

The constellation of Cancer represents the giant crab that attacked Hercules during the second of the 12 labors he performed as penance for killing his family. It was sent by the jealous goddess Hera to thwart Hercules as he battled the water serpent Hydra, but he killed it with his club.

The Zodiac sign Cancer is ruled by the Moon. People born underneath this star sign are characterized being emotional, nurturing, highly intuitive, as well as sensitive and at times insecure.

ELEMENT: Water

BODY: Chest, breasts and stomach.

GENERAL: Comfort, self care and maternal energy.

Constellation: Cancer

(Brittanica)

Can be seen...

Southern Hemisphere: *December - February.*

Northern Hemisphere: *March - July.*

Notable Facts:

Cancer is Latin for "the crab." It wields the Beehive
Cluster, a open star cluster that can be seen with the naked
eye, aka M44.

Cancer is relatively faint. In China, this constellation was
associated with the "Chariot" constellation.

LEO

In mythology, Leo represents the Nemean lion. **The hero Heracles (or Hercules) killed the Nemean lion as part of a series of tasks he had to perform.** The lion's golden mane protected it from any assault, its claws were sharper than any weapon forged by humanity.

The Zodiac sign Leo is ruled by the Sun. People born underneath this star sign are characterized as slightly temperamental but highly loyal and loving towards their inner circle. Leos love luxury, they are generous, ambitious and love attention.

ELEMENT: Fire

BODY: Heart

GENERAL: Stability, loyalty, consistency and leadership.

Constellation: Leo

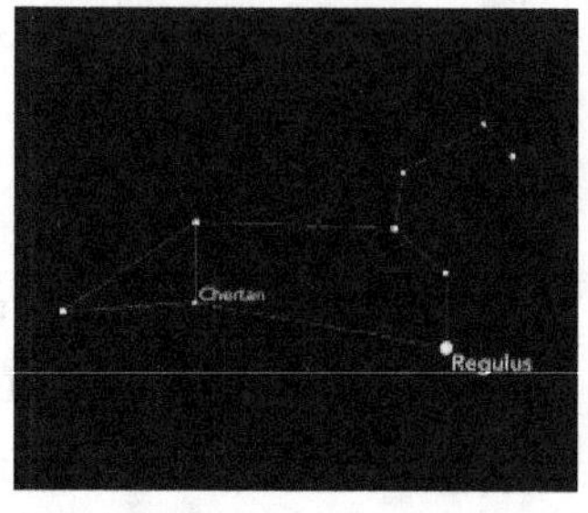

(Star Registration)

Can be seen...

Southern Hemisphere: *March - May.*

Northern Hemisphere: *February - May.*

Notable Facts:

Ursa Major is to the north of this constellation. It's brightest star is Regulus (Alpha Leonis), a blue-white main-sequence star with a magnitude of 1.35.

The Leonids meteor shower occurs annually in November, associated with the comet Tempel-Tuttle. There can sometimes be meteor storms.

VIRGO

Most myths view Virgo as a virgin maiden that carries her associations with wheat. In Greek and Roman mythology, the constellation is related to Demeter, the Greek goddess of the harvest, or her daughter Persephone, queen of the Underworld.

The Zodiac sign Virgo is ruled by Mercury. People born underneath this star sign are characterized as logical, practical, and systematic in their approach to life. This earth sign is a perfectionist at heart and isn't afraid to improve skills through diligent and consistent practice.

ELEMENT: Earth

BODY: Digestive system, pancreas, small intestines, eyes, and ears.

GENERAL: Detail, organization, aloof and critical.

Constellation: Virgo

(Star Registration)

Can be seen...

Southern Hemisphere: *March - May.*

Northern Hemisphere: *March - September.*

Notable Facts:

Virgo is Latin for "the virgin." Within this constellation is the Virgo Cluster, a large cluster of galaxies including Messier 87 (M87), a supergiant elliptical galaxy with a supermassive black hole.

The asterism known as the "Y of Virgo" is formed by the stars Spica, Porrima and others. The Virginids meteor shower occurs annually in late April.

LIBRA

Libra is related to the Greek Goddess of Justice, **Themis**, whose daughter, Astraea went up to heaven and became the constellation of Virgo. Both goddesses carried the scales of justice, which became the symbol for Libra.

The Zodiac sign Libra is ruled by Venus. People born underneath this star sign are characterized as extroverted, cozy, and friendly people. They like attaining balance, harmony, peace, and justice in the world. With their vast stores of charm, intelligence, frankness, persuasion, and seamless connectivity, they are well-equipped to do so

ELEMENT: Air

BODY: Kidneys, butt, skin and lower back.

GENERAL: Calm, balance, justice and easy-going.

Constellation: Libra

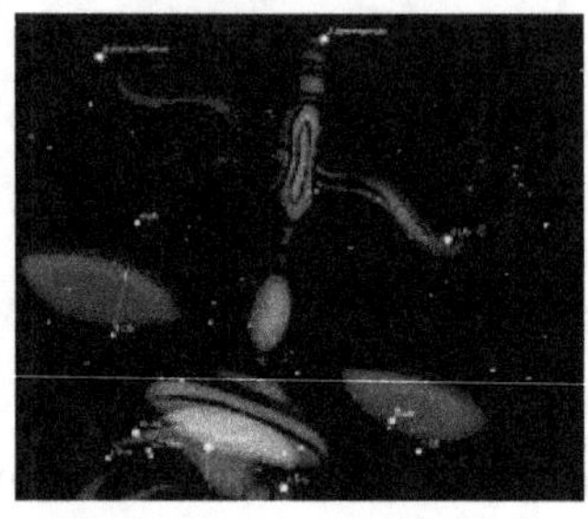

(Space)

Can be seen...

Southern Hemisphere: *March - May.*

Northern Hemisphere: *April - July.*

Notable Facts:

Libra is Latin for "the scales." Centaurus can be found to the south of this constellation. Deep-Sky objects include Messier 4 and 19; two globular clusters.

It's brightest star is Zubenelgenubi (Alpha Librae), a binary star system. In Egyptian culture, Libra was associated with Ma'at, who represented truth and justice. The scales were used to weigh the hearts of the deceased against a feather to determine their afterlife fate.

The Lyrids meteor shower occurs in April.

SCORPIO

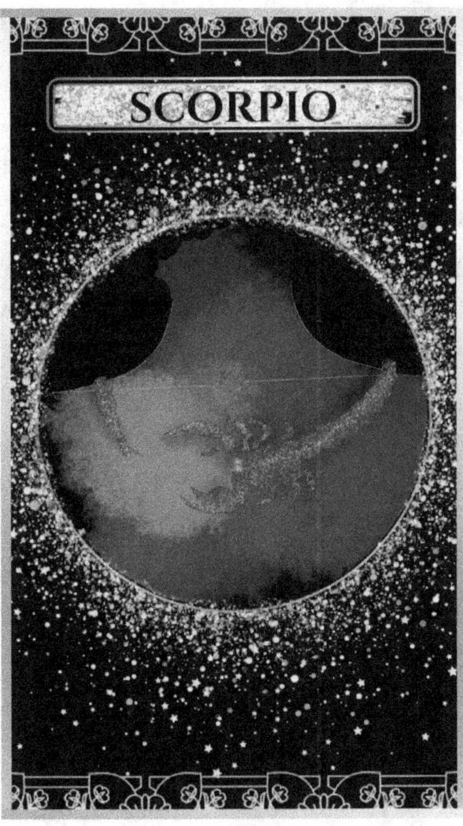

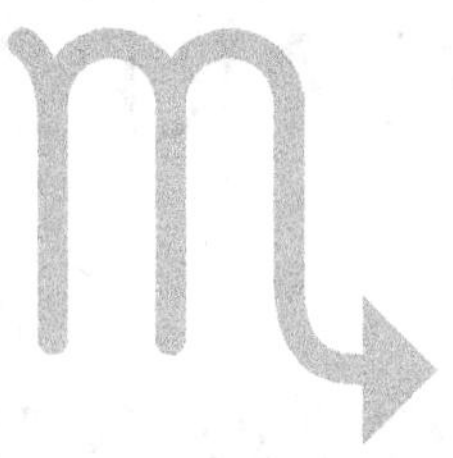

Skorpios (Scorpius) was a giant scorpion sent by Gaia the Earth to slay the giant Orion when he threatened to slay all the beasts of the world. Orion and the Scorpion were afterwards placed amongst the stars as the constellations of the same name.

The Zodiac sign Scorpio is ruled by Pluto. People born underneath this star sign are characterized as extremely emotional and craving intimacy. They have a powerful presence and demanding personalities and an all around mysteriousness.

ELEMENT: Water

BODY: Hips and reproductive organs.

GENERAL: Passion, excitement, living life to the fullest, reproduction, fierceness and destruction.

Constellation: Scorpio

(Space)

Can be seen...

Southern Hemisphere: _June - August._

Northern Hemisphere: _June - September._

Notable Facts:

Antares (Alpha Scorpii) is the brightest star. Shaula
(Lambda Scorpii) and Lesath (Upsilon Scorpii) represent
the scorpion's stinger.

Messier 4 and 6 can be found here. Messier 6 is the
Butterfly Cluster, an open star cluster known for its unique
shape.

SAGITTARIUS

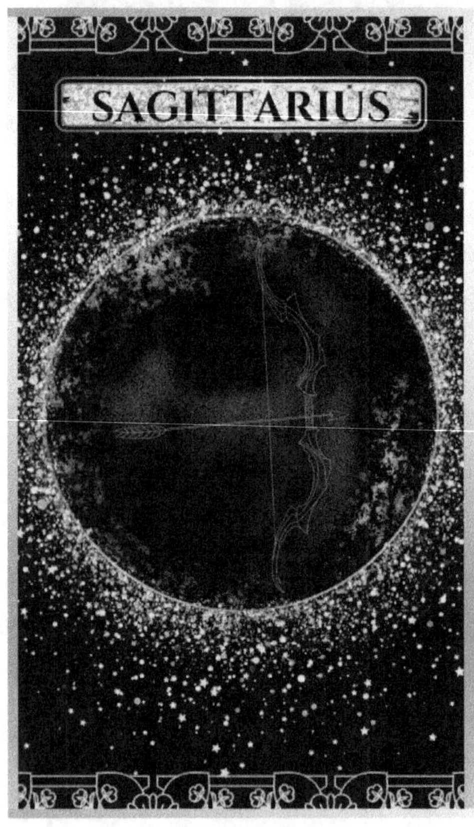

Sagittarius is commonly thought to represent a centaur, a war-like creature with the torso of a man and the body of a horse. Sagittarius is most often associated with Crotus, the son of Pan (the goat-god) and Eupheme (the Muses' nurse).

The Zodiac sign Sagittarius is ruled by Jupiter. People born underneath this star sign are characterized as generous, idealistic and having a great sense of humor. They are known to promise more than they can deliver, having impatience and will say anything no matter how undiplomatic.

ELEMENT: Fire

BODY: Hips, thighs, legs, lower back, and pelvis

GENERAL: Adaptability, flexibility and freedom.

Constellation: Sagittarius

(Star Registration)

Can be seen...

Southern Hemisphere: *June - August.*

Northern Hemisphere: *June - September.*

Notable Facts:

Sagittarius is Latin for "the archer." Kaus Australis (Epsilon Sagittarii) is the brightest (blue-white) star. Kaus Media (Delta Sagittarii) is part of the "Teapot" asterism.

Messier 8 (Lagoon Nebula), Messier 20 (Trifid Nebula) and Messier 22 (globular cluster) are found here. This constellation is home to the Galactic Center of the Milky Way galaxy, near the star Sagittarius A which is rich in stars, gas and dust...along with a supermassive black hole.

OPHIUCHUS

Ophiuchus is also known as the "serpent bearer". The constellation "holds" Serpens, who's venom can either kill or cure. This constellation has sometimes been referred to as the 13th zodiac sign.

The myth associated with this constellation/sign is about Asclepius, the son of Apollo, who was able to bring people back to life with his healing powers. He learned to do this after seeing one snake bring healing herbs to the other.

GENERAL: Healing, medicine, reverse death, necromancy, intellectual, doctors and odds in your power.

Constellation: Ophiuchus

(Star Name Registry.)

Can be seen...

Southern Hemisphere: *December - March.*

Northern Hemisphere: *June - September.*

Notable Facts:

Ophiuchus has Hercules to the north of it and Serpens to the west. Ras Alhague (Alpha Ophiuchi) is the brightest star - a binary star system.

It contains globular clusters such as Messier 10, 12 and NGC 6293. This constellation is the 11th largest. The Ophiuchid meteor shower occurs in late December and is associated with the comet C/1861 G1 Thatcher.

CAPRICORN

In Greek mythology, the god Pan was transformed into a half-goat, half-fish when he dived into the Nile River to escape the giant Typhon. He ruled over forests and woodlands, flocks and shepherds.

The Zodiac sign Capricorn is ruled by Saturn. People born underneath this star sign are characterized as workaholics, dedicated, ambitious, goal-oriented, serious and resilient. They can also be very spiritual as well as leaders in religious institutions.

ELEMENT: Earth

BODY: Knees, joints, skeletal system and teeth.

GENERAL: Mischievous, troublemaker, independence, inflexibility, stubborn and working towards money.

Constellation: Capricorn

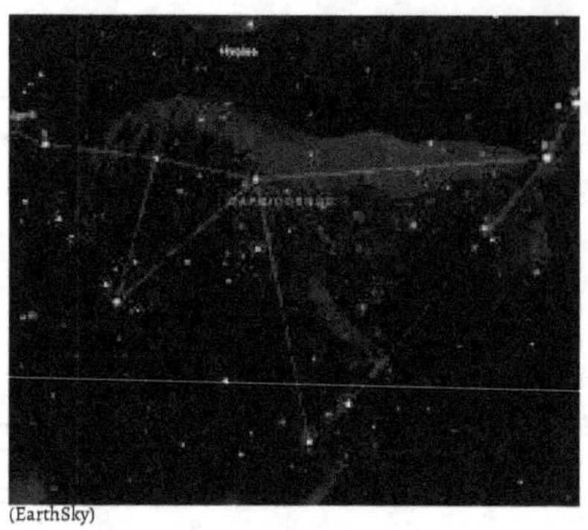
(EarthSky)

Can be seen...

Southern Hemisphere: *June - August.*

Northern Hemisphere: *July - September.*

Notable Facts:

Microscopium is north to this constellation. The brightest star is Dabih (Beta Capricorni), a binary star system.

Messier 30 and NGC 6934 are globular clusters that find their homes here.

The Capricornids meteor shower occurs in late July.

AQUARIUS

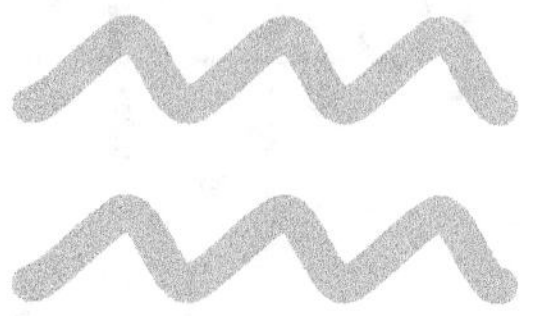

Aquarius is sometimes associated with Deucalion, the son of Prometheus who built a ship with his wife Pyrrha to survive an imminent flood. They sailed for nine days before washing ashore on Mount Parnassus.

The Zodiac sign Aquarius is ruled by Uranus. People born underneath this star sign are characterized as advanced, self-reliant, clever, exceptional, and optimistic. They lack a solid form and refuse to be categorized.

ELEMENT: Air

BODY: Shins, calves, ankles and circulatory system.

GENERAL: New radical beginnings, thinking outside of the box and flow of ideas.

Constellation: Aquarius

(Star Name Registry.)

Can be seen...

Southern Hemisphere: *November - February.*

Northern Hemisphere: *September - January.*

Notable Facts:

Eta Aquariids meteor shower occurs annually in early May and is associated with Halley's Comet. Aquarius is Latin for "the water bearer." Pegasus can be found to the west of this constellation.

Messier 2 (globular cluster) and Messier 76 ((Little Dumbbell Nebula) planetary nebula) are found in this region.

PISCES

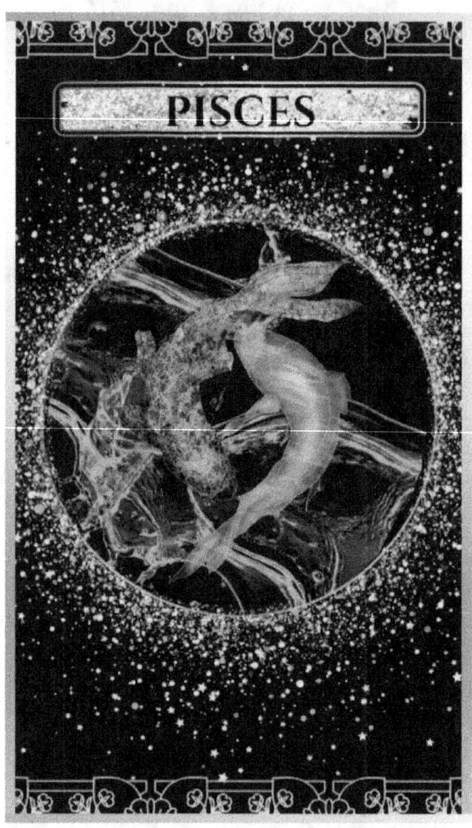

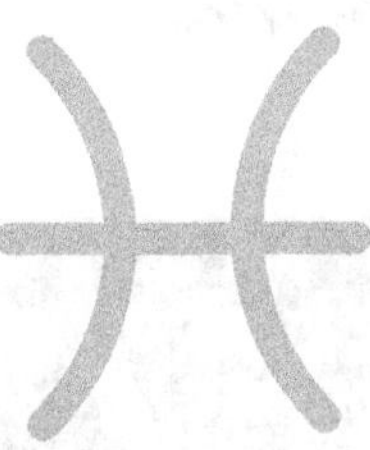

Pisces represents two fish with their tails tied together by ribbons. According to Greek mythology, the fish represent the goddess Aphrodite and her son, Eros. They are said to have transformed into fish and plunged into the Euphrates River to escape the fearsome monster Typhon.

The Zodiac sign Pisces is ruled by Neptune. People born underneath this star sign are characterized as highly creative and imaginative. They are in tune with their and others emotions, worrying a lot about the effect that their actions might have on others.

ELEMENT: Water

BODY: Hands and feet

GENERAL: Love, devotion, fantasy and pleasure.

Constellation: Pisces

(Star Name Registry.)

Can be seen...

Southern Hemisphere: *March - May.*

Northern Hemisphere: *September - February.*

Notable Facts:

Pisces is Latin for "the fish." The brightest star is Alrescha (Alpha Piscium), a binary star system. Messier 74 (face-on spiral galaxy/the Phantom Galaxy) and NGC 247 (a barred spiral galaxy part of the Sculptor Group) are within it's region.

The Pi Puppids meteor shower occurs in late April. The "Circlet" asterism is a group of stars representing the head of one of the two fish.

Sky Maps

The Night Sky

Here you will find various sky maps of the constellations above. These serve as useful tools to use in real life and for visualization.

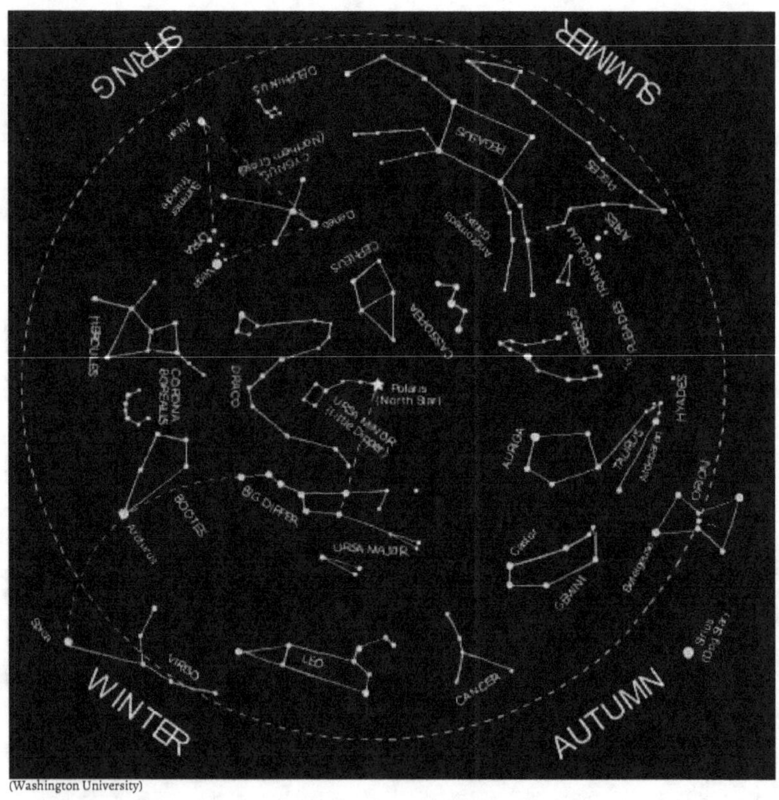

(Washington University)

(In the Sky)

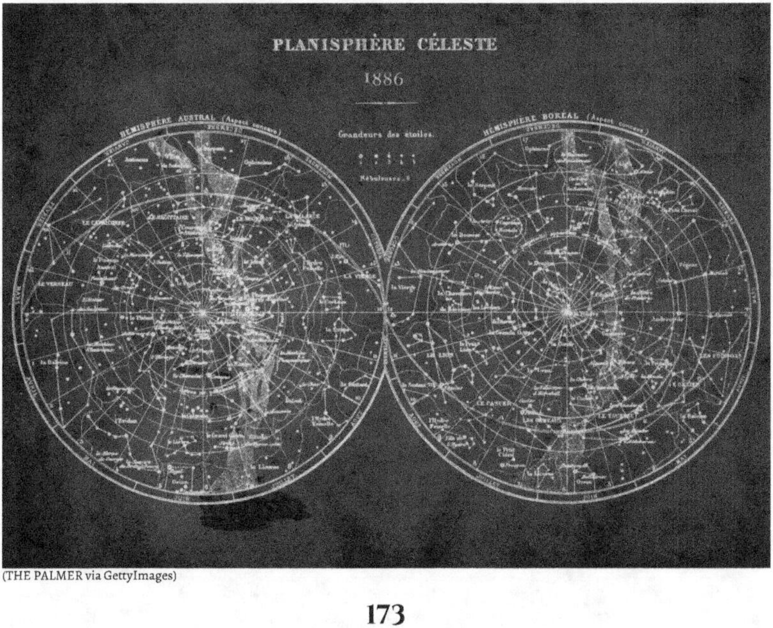

(THE PALMER via GettyImages)

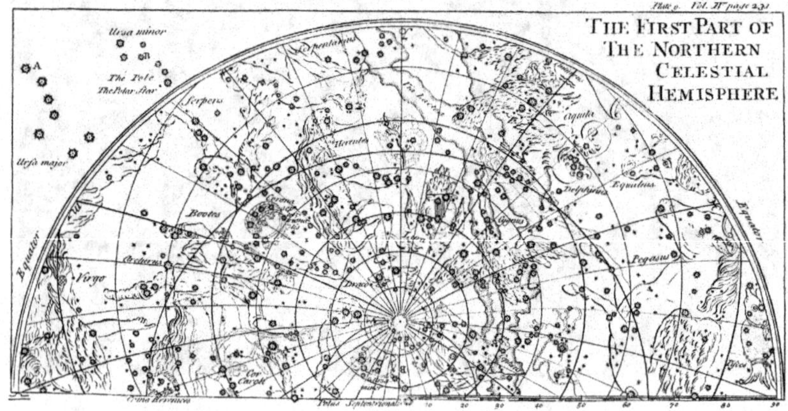

(Photos.com)

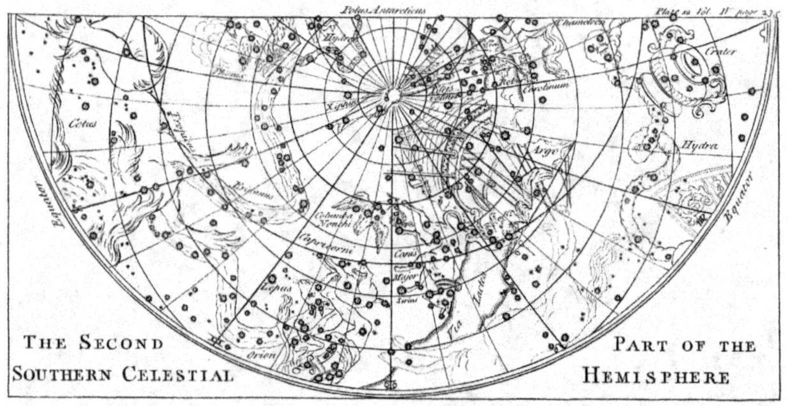

(Photos.com)

Resources

Star Walk 2 (phone app): For live constellations based on where you are currently located.

Heavens-Above (web and mobile): shows live maps of the sky based on your time and location. Focuses on satellites.

Celestia (computer): provides real-time 3D simulations of the solar system and beyond. You can travel through space and time to see celestial events, planets and more from various points of view.

Stellarium (computer): shows the sky in a realistic 3D format as you would with the naked eye, binoculars or telescope.

NASA & the James Webb Telescope provide photos and information online free to the public all the time.

Moon Calendar (app): shows the phases of the moon, the current sign as well as potential affects based on astrology.

As you may have noticed, not all constellations are listed prior to this page. In total there are 88 modern constellations. Before this, Ptolemy listed 48 constellations, some of which were abbreviated or broken into smaller constellations. Across various cultures, they also translated the stars in multiple ways.

With all of this in mind, there are endless ways to enjoy the night sky. With this solid foundation, along with the other resources (and many more which can be found online), you can feel free to customize your own experience!

The Specter Stargazer is proud to provide you with a good baseline of knowledge, as well as a more in-depth tool for those who purchased a copy of the Specter Cosmos Tarot.

I hope the discoveries within this work aid you in your spiritual and/or astronomy journey! Most importantly - have fun! The world is a beautiful thing and it should be adored. Let's bring back the lost art of stargazing, one page at a time.

Citations

NASA. "A Deep Field Image from the Hubble Space Telescope." *NASA Science*, 30 Sept. 2008, https://science.nasa.gov/image-detail/amf-pia11351/.

NASA. "Mercury." *NASA*, 2023, https://science.nasa.gov/wp-content/uploads/2023/04/EW0108829708G4release_mercury-jpg.webp.

NASA. "Mercury." *NASA*, 2023, https://science.nasa.gov/wp-content/uploads/2023/05/pia19423-mercury-jpg.webp.

NASA. "Venus." *NASA*, 2023, https://science.nasa.gov/wp-content/uploads/2023/05/venus-single.png.

NASA. "Venus from Parker Solar Probe." *NASA*, 2023, https://science.nasa.gov/wp-content/uploads/2023/05/venus-from-parker-solar-probe.webp.

NASA. "Image of the Moon." *NASA*, n.d., https://images-assets.nasa.gov/image/PIA01544/PIA01544~orig.jpg?w=1708&h=248&fit=clip&crop=faces%2Cfocalpoint.

NASA. "Mars Perseverance Sol 1282: Right Navigation Camera (Navcam)." *NASA*, 28 Sept. 2024. https://mars.nasa.gov/mars2020/multimedia/raw-images/NRF_1282_0780762783_972ECM_N0600390NCAM00709_09_095J

NASA. "A simulated view of Mars as it would be seen from the Mars Global Surveyor spacecraft." *NASA*, 2 Mar. 2021. https://unsplash.com/photos/mars-on-a-black-background-N3BQHYOVq5E

NASA. "Top View Land Under Cloud..." *NASA*, 26 Dec. 2015. https://unsplash.com/photos/top-view-land-under-clouds-OodEH-UPj68

NASA. "Hubble Provides Unique Ultraviolet View of Jupiter." *NASA*, 03 Nov. 2023. https://www.nasa.gov/image-article/hubble-provides-unique-ultraviolet-view-of-jupiter/

NASA. "Jupiter Swirling Storms." NASA, 27 Oct. 2021.
https://photojournal.jpl.nasa.gov/catalog/PIA24971

NASA. "Jets at Jupiter." NASA, 27 Oct. 2021. https://science.nasa.gov/image-detail/amf-pia24964/

NASA. "Saturn Atmospheric Changes." NASA, 28 Mar. 2001.
https://science.nasa.gov/image-detail/amf-pia03152/

NASA. "Saturn in Full View." NASA, 2 Aug. 2004,
https://science.nasa.gov/image-detail/amf-pia05425/
NASA. "Northern Hemisphere of Saturn." NASA, 20 Dec. 1999,
https://science.nasa.gov/image-detail/amf-pia02230/

NASA. "GIF of Uranus' Magnetic Field." NASA, 24 Mar. 2020,
https://science.nasa.gov/image-detail/amf-pia02230/

Webb Space Telescope. "Uranus Close-up." Webb Space Telescope, 18 Dec. 2023,
https://webbtelescope.org/contents/news-releases/2023/news-2023-150

NASA. "NASA's Webb Scores Another Ringed World With New Image of
Uranus." NASA, 6 Apr. 2023, https://www.nasa.gov/solar-system/nasas-webb-scores-another-ringed-world-with-new-image-of-uranus/

NASA. "Neptune False Color Image of Haze." NASA, 29 Jan. 1996,
https://science.nasa.gov/image-detail/amf-pia00057/

NASA. "Neptune Scooter." NASA, 8 Jan. 1998, https://science.nasa.gov/image-detail/amf-pia01142/

NASA. "Crescents of Neptune and Triton." NASA, 23 Jan. 1999,
https://science.nasa.gov/image-detail/amf-pia01142/

NASA. "Color Image of Pluto." NASA, 13 Jul. 2015,
https://science.nasa.gov/image-detail/color-image-of-pluto-pia20291-2/

NASA. "Pluto Haze." NASA, 17 Mar. 2016, https://science.nasa.gov/image-detail/amf-pia20536/

NASA. "Pluto Color Map." NASA, 19 Jan. 2017, https://science.nasa.gov/image-detail/amf-pia11707/

Author Biography

Jessica Specter is a serial-author, bear conservationist, artist and marketing professional. She received a Bachelor's degree in Communication from the University of North Florida in April of 2021. In late 2021 she would start freelancing creative services under Lavender Moon Media, LLC. In July of 2023 she established her bear conservation nonprofit, which was later 501(c)3 approved in 11/2023.

Within the past three years she has created multiple original stories, over 350 pieces of original artwork, two tarot decks (Specter Gold Tarot (completed on 04/2023) and Specter Cosmos Tarot (completed on 07/2023,) and other products/services online - even including personal tarot readings.

In the future, she plans to create another book as her first full-length fiction novel.

Links: linktr.ee/jessicaspecter